THE AIR PILOT'S SATELLITE POSITIONING SYSTEMS

THE AIR PILOT'S GUIDE TO
SATELLITE POSITIONING SYSTEMS

WALTER BLANCHARD, FRIN

Airlife
England

Copyright © 1995 by Walter Blanchard

First published in the UK in 1995
by Airlife Publishing Ltd

British Library Cataloguing in Publication Data
A catalogue record for this book
is available from the British Library

ISBN 1 85310 599 6

All rights reserved. No part of this book may be reproduced or
transmitted in any form or by any means, electronic or mechanical
including photocopying, recording or by any information storage
and retrieval system, withour permission from the Publisher in writing.

Typeset by Servis Filmsetting Ltd, Manchester
Printed in England by Livesey Ltd., Shrewsbury.

Airlife Publishing Ltd
101 Longden Road, Shrewsbury SY3 9EB, England

Contents

Foreword   9

Introduction: The Background to Radio-positioning Systems.   13

Chapter 1.
    Position-Finding for Navigation.
        1. Defining Position   15
        2. Position in Latitude and Longitude   16
        3. The Geometry of Two-dimensional Fixing   17
        4. The Derivation of Lines of Position   22
        5. Positioning Systems   23
        6. Three-dimensional Effects   23

Chapter 2.
    Basic Satnav Systems.
        1. Satellite Orbits   25
        2. How Many Satellites?   31
        3. Measuring Distance to Satellites   34
        4. Dilution of Precision (DOP)   43
        5. Determination of Satellite Position   44
        6. Satellite Timing   46
        7. Satellite Management Data   47
        8. Data Transmission to the User   48
        9. Integration with Ground-based Systems   48
        10. Accuracy   48

Chapter 3.
    The Global Positioning System – GPS/Navstar.
        1. The Defense Navigation Satellite System   58
        2. Launch History   60
        3. System Operation   65
        4. The Operational Control System   69
        5. GPS Errors and Selective Availability   72
        6. Differential GPS   75

7.	GPS Integrity	75
8.	Reliability	75

Chapter 4.
Aerials and Receivers for GPS.
1.	The Aerial	76
2.	Receivers	79

Chapter 5.
Differential Satnav.
1.	Advantages of DGPS	85
2.	DGPS System Components	87
3.	DGPS Error Sources	94
4.	Area of DGPS Coverage	97

Chapter 6.
The Global Navigation Satellite System – GLONASS.
1.	System Description	99
2.	Current Status	104
3.	Prospects for Deployment	105

Chapter 7.
Other Satellite Positioning Systems.
1.	Inmarsat	107
2.	Geostar/Locstar	110
3.	Omnitracs/Euteltracs	111
4.	Iridium	114
5.	Orbcomm	114
6.	Odyssey	116

Chapter 8.
Applications of Satnav for Aviation.
1.	Integrity, Availability and Augmentation	118
2.	Airfield Approach and Landing	124
3.	Combined Integrity and Augmentation Systems	127

Appendix 1.
    Methods of Obtaining Range by Radio   130
Appendix 2.
    How Satellite Positions In Space are Defined   137
Appendix 3.
    The Transit System   143

Appendix 4.
    Spheroids and Datums   147
Appendix 5.
    Ordnance Survey Maps and Mapping Policy   149
Appendix 6.
    International Frequency Allocations for Navigation   155
Appendix 7.
    The Spread-Spectrum Modulation Method   159
Appendix 8.
    Extract from the U.S. Federal Radionavigation Plan 1992   164
Appendix 9.
    Practical DGPS Systems   167
Appendix 10.
    List of GLONASS Satellite Launches   171
Appendix 11.
    Comparison of GPS and GLONASS   174
Appendix 12.
    The Wide-Area Augmentation System WAAS   175
Appendix 13.
    The European Geostationary Navigation Overlay Service – EGNOS   183
Appendix 14.
    Institutions Involved in the Operation of Aviation Navaids   186

References and Sources of Information.   191
Index   194

Foreword

In 1887 Heinrich Hertz built the world's first radio transmitter and received its transmissions on a loop of wire coupled to a simple detector. He noticed that if he turned the loop the strength of the signal changed and whether he realised it or not, he thereby became the first person ever to take a radio bearing. He probably did not appreciate the navigational significance of what he had done, but it does give some basis for claiming that the navigational properties of radio waves were observed before those of communication.

Many attempts to use radio techniques for obtaining position followed. As usual with any new technical development, techniques and equipment were not at first good enough to provide worthwhile accuracy, but even when they later improved Mother Nature stepped in and ruined accuracy with skywaves. A positional accuracy of 300 metres needs timing measurements of better than one-millionth of a second and quite minor propagation problems introduce this sort of error. Engineering techniques improved as time went on, but the physical laws of the universe did not change. There is no point using incredibly accurate timing techniques when propagation introduces errors thousands of times larger; the only solution is to change to a different mode of propagation. Unfortunately, the line-of-sight propagation that is least likely to suffer errors offers only short ranges, even for aircraft, and it was not until the advent of satellites that consistently high accuracy over huge areas became possible.

The magnitude of this revolution in radio navaids is still only dimly discernible some twenty-five years after the first satnav system, Transit, was put into operation. Satnav prophets gazing into their crystal balls (or rather, VDU screens!) are making startling predictions about future satnav and its uses. Perhaps even more startling is the speed with which they are coming about. It is only fifteen years since it was predicted, to some scepticism, that a wristwatch position-finder would be with us by the turn of the century, but in fact one was announced by a very reputable manufacturer in 1994.

The benefits of having instant and accurate position available everywhere are incalculable. No doubt, there will also be misuse but it would be a major mistake to deny it on those grounds. No-one seriously proposes that the telephone service should be withdrawn because it is sometimes misused.

The prime examples of satellite positioning systems are of course the

The Air Pilot's Guide to Satellite Positioning Systems

Fig 1 – The JRC Wristwatch GPS receiver

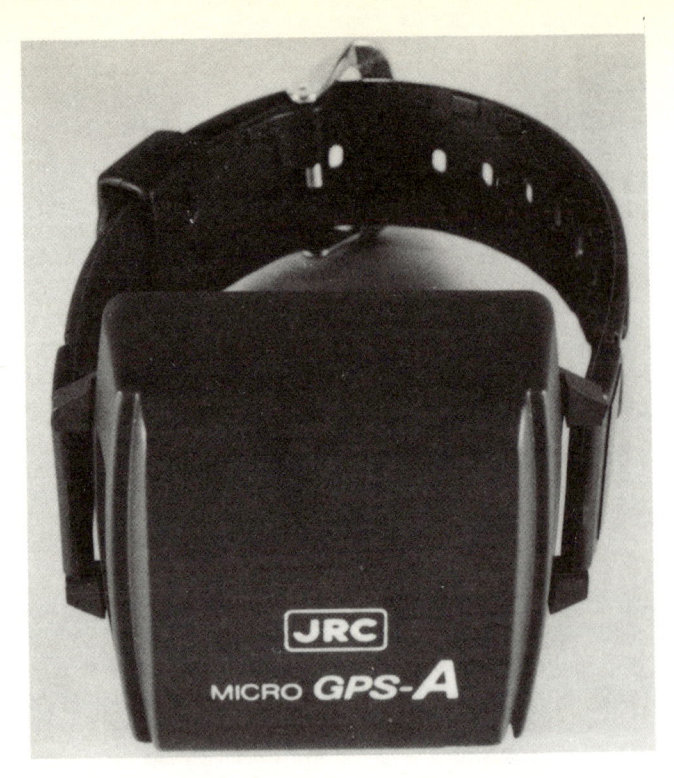

American Global Positioning System (GPS) and Russia's GLONASS. Outline descriptions of GPS can be found in any GPS receiver manufacturer's handbook, which, by the way, is where to look if you simply want to know how to operate a new GPS set. This book does not give receiver operating instructions – manufacturers know their own equipment better than anyone else and any attempt to cover all the sets now available would be tedious and repetitive even if possible. But GPS and GLONASS are not all there is to satnav – in 1994 there were twenty-two other satnav systems actual or proposed. Not all of them will become operational, in fact probably rather few, but the basic principles and requirements for satnav remain the same no matter which system is involved.

The aim of this book is to describe how satnav systems are designed and work, especially from the aviation viewpoint, without getting too heavily involved in mathematics and electronic engineering. In this digital age, most people's brains still seem to work best in an analogue fashion and I have therefore attempted to give analogue explanations wherever possible to prevent that feeling of lassitude on encountering page after page of solid mathematics! However, it does assume some knowledge of elementary maths, physics and electronics and a nodding acquaintance with navigational terminology. It is written for the interested satnav user who would like to go into things more

Foreword

deeply than simply knowing which knob to twiddle, and also to provide technical background material for the increasing number of non-specialists having to deal with GNSS and its implementation. Some of the more abstruse material has been put into appendices and references have been cited to facilitate further reading if desired.

Because not everyone who reads this book will be familiar with navigational principles, it starts with a short description of basic navigational position-finding techniques. Even qualified navigators, blinded by new technology, occasionally seem to forget what it is satnav actually has to do! GPS, differential GPS and GLONASS are described in some detail and commercial systems like Euteltracs, Iridium and Starfix are included. The problems encountered in using satellite positioning technology for aviation purposes are described, and there are sections on integrity, monitoring by ground stations (GAIT), built into the receiver (RAIM), and on augmentations like the American WAAS.

I have fallen in with popular practice and used the abbreviation 'satnav' but it must be emphasized that satellite positioning systems on their own are not navigation systems. GPS acknowledges this by its full title 'The Global Positioning System', which makes it clear that its function is to provide position and not to navigate. Navigation comprises far more than just determining position, important as that is, but that is another matter and outside the scope of this volume!

Introduction

The Background to Radio-positioning Systems

Someone once said that 'history is bunk' but we should also remember that other popular saying – 'There's nothing new under the Sun'! Satnav seems to be a marvellous new invention and so it is in many respects, but it can also be seen as the culmination of developments that have taken place over a long span of years. It could not have been developed without the knowledge obtained from its predecessor land-based systems; particularly in how they are implemented and operated.

Notwithstanding Hertz's loop aerial, the earliest real use of radio for navigational purposes was the transmission of time signals for the calibration of clocks on board ship. Marine navigators had little difficulty measuring latitude using their sextants but longitude depended on how accurately time could be kept. Wireless time signals were first broadcast by the US Navy from New York in 1904, followed by Germany in 1907, France in 1910, and the UK in 1927.

Coming to radio-positioning, as early as 1900 Marconi applied for a patent on what was probably the first radio device designed specifically to help position-finding. It was actually only a very long piece of wire, that because of its great length (several miles) would produce a 'beam' in the direction it was pointing if it was used as an aerial for a transmitter. The idea was that a ship sailing through this 'beam' would be able to get a rough idea of where it was relative to the transmitter. Unfortunately the 'beam' was so broad it was navigationally almost useless, and if the ship was stationary it was no help at all.

Then, in 1904, Fessenden patented in the USA what would have been the first area-coverage radio positioning system if it could have been made to work. It was a method of obtaining range measurements from a number of transmitters which, by using range instead of bearings would overcome the loss of accuracy of bearings, at increasing distances. Ranges were to be obtained by measuring signal strength, since everyone knew radio signals got weaker as distance increased. The effects of sky waves were not then known and he thought that signal strength could be exactly calibrated.

The first operational use of radio for navigational position-fixing was by a German naval airship which attacked England in April 1915 using a radio direction-finding service operated by the German Navy, and by the end of World War 1 considerable development had been done on direction-finding.

The French experimenter Mottez, in 1922, was the first to see the possibilities of using the hyperbolic principle in connection with area-coverage radio-positioning systems. He noted that these systems would have an inherent problem with ambiguity and suggested that every so often the transmitters would change to a longer wavelength that would produce fewer ambiguities although of less accuracy. This idea reached fruition in a modified form in Decca Navigator, Loran and other systems many years later, but even in GPS rather similar problems arise when attempts are made to increase its accuracy by using phase and cycle tracking and the solution is basically the same as Mottez proposed.

Thus, by 1922 virtually all the fundamental principles of radio-positioning systems had been stated. Many systems of ever-increasing accuracy and range were subsequently developed, but all were limited by propagation disturbances. At the lower frequencies in particular, sky waves, thunder-storms, rain, static and varying speeds of propagation over differing surfaces all had their effects and limited accuracy no matter how good the basic engineering was. Higher frequencies, although not suffering so much from these disturbances, had only short range. Designers had to make a choice between very high frequencies – giving high accuracies but only short ranges – or lower frequencies for better ranges but lower accuracies.

There was no way of breaking out of this circle until satellites came along, but when they did the difference was startling. Very high accuracy obtained through the use of purely line-of-sight propagation is one of the two major advantages satellites offer for radio-positioning systems; the other is of course the enormous radio ranges possible from the height of a satellite. But nothing comes free, and satellites are no exception. Apart from their sheer cost, the very nature of their world-wide coverage and universal applicability demands new methods of international control and operation no longer based only on the local or national considerations that were all that were necessary for limited-range Earth-based systems. In the development of a civil Global Navigation Satellite System (GNSS), they may be severe enough to be more of a problem than technicalities.

Chapter 1

Position-Finding for Navigation

1. Defining Position

Position is always relative – it does not exist independently. Try telling someone where you are without referring to any other object! Position can only be given as a displacement from another object or point whose position is itself already known. And, of course, that position in turn is only relative to yet something else. It is a matter of definition where to stop in this chain, but wherever it is is then called the reference 'datum' or 'origin'. Of course it would be silly to push this too far and since we Earth-bound navigators are only interested in where we are on Earth the final reference datum is usually some geographical feature on the Earth itself, or an artifical grid superimposed on it.

The simplest way of defining position is to give it as a distance (range) and bearing from some known object, e.g. 'I am two miles bearing 060° from the XXX VOR/DME' (Fig 2). In this case, the VOR/DME is the reference datum for the fix and it is assumed that everyone knows where it is.

A simpler statement such as 'I am two miles from the XXX VOR/DME' does not define position, only that you are somewhere on a circle of two miles radius from it. Similarly, 'I am bearing 060° from the XXX VOR/DME' – you could be anywhere from a few miles to hundreds of miles away. So it is important to note that at least two lines of position are always needed to give position but note also (and it is important for satnav) that it is also assumed in all this, that vertical position (altitude) does not matter. We are thinking of a flat map – a flat Earth – and the transmitters and users all being on its surface.

Range and bearing lines drawn around a beacon form a grid and if drawn on an ordinary map will be superimposed on coastlines and other geographical features. It is then easy to turn range and bearing from the beacon into range and bearing from any other point on the map by a little simple protractor-and-divider work. The map is in effect acting as an analogue computer for turning position related to one point into position related to another. Now that we have digital computers the need to actually do this drafting work has disappeared, and just as well in a small cockpit!

For some types of navigation that's all you need. If you are trying to avoid overflying Heathrow, then all you want to know is where it is and where you

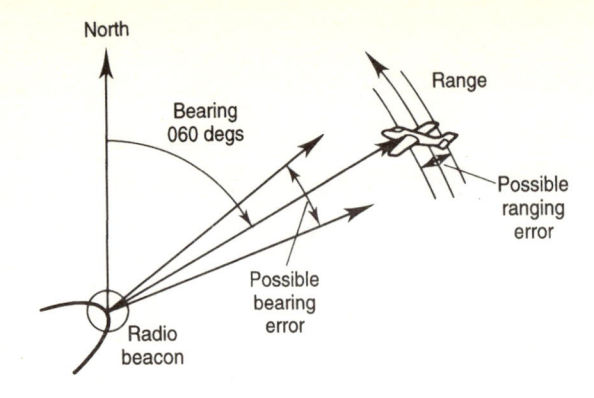

Fig 2 – Range/Bearing Fix

are relative to it. But it would be quite different trying to navigate across an ocean or desert out of sight of any definable point for hours on end.

2. Position in Latitude and Longitude

Another example of a grid is latitude and longitude, designed to be used anywhere in the World. Superficially it appears to provide a means for defining position without relating to any known point, but not so. The lat/lon grid uses the North and South Poles (or, to be pedantic, the 'axis of revolution of the Earth') as its datum and all it does is to give position relative to those Poles. If the Poles moved then the whole lat/lon grid would also move (actually, they do, slightly, so a mean Polar position is adopted). A position quoted in lat/lon is in effect giving range (latitude) and bearing (longitude) to the nearest Pole. Always remember that:

> 1. Lat/lon is not a stand-alone reference system that works independently; it is only one form of artificial grid.
> 2. Lat/lon is a two-dimensional system – it exists only on the surface of whatever geoid ('geoid' = average shape and size of the Earth) is used to represent the Earth. So *it is different for different geoids – and there are many!*
> 3. Lat/lon is a position transponder between its own origin and the user's position. If its origin is not properly defined then the resultant lat/lon will be in error no matter how accurate electronic measurements may be.

Because of this two-dimensional limitation, lat/lon – on its own – is inadequate for specifying positions derived from satellites, which are three dimensional by nature. Only if the third dimension, usually rather loosely called 'height', is already known can lat/lon be used for the other two. However, what 'height' really means is a complicated matter and is discussed more fully in Chapter 2.

To determine 'height' requires that, as for lat/lon, the geoid is closely defined in shape, size and position. If it is defined incorrectly then all distances and heights will be wrong. And, of course, a position given using one geoid will not be correct on a map drawn using another one. There are quite a large number of different geoids and it is necessary to be very careful about which one is being used. Ideally it would always be specified by quoting, for instance, '5127N 0534W (WGS–84)', 'WGS–84' denoting the geoid. But there is no need to get paranoic over geoids. The differences between them over, say the UK, are really rather small (100–150 metres) and only become important when really high accuracy is needed. A more detailed description of geoids with a listing of the better-known ones is given in Appendix 4.

We will not here go into the fourth dimension – time – which is just as important if the user is in a vehicle travelling at high speed. A 600-knot aircraft is travelling at about 300 metres per second and a fix mis-timed by only one second would make nonsense of all the claims to accuracy of any satnav (or, for that matter, any other system).

These points have been stressed because they are much more important for satellite positioning than they were for land-based systems, as will become even more obvious later.

3. The Geometry of Two-dimensional Fixing

Ignoring three-dimensional aspects for a moment and considering only two dimensions, it is not always possible to obtain a simultaneous range and bearing from a single point, and two different sources must be used. But two bearings or ranges require two geographically-separate sources if they are to be measured simultaneously and a new problem is then introduced, that of geometry. Fig 3 illustrates the point; two bearings (it would be similar for two ranges) have been obtained from two sources but where is the fix?

The angle of intersection of the two position lines is zero, so there is no crossing point. If they had crossed even at quite an acute angle then the problem might be solved, although it would depend on the precision of the measurement. No position line can be measured absolutely pre-

Fig 3 – Fix Sources 180° Apart

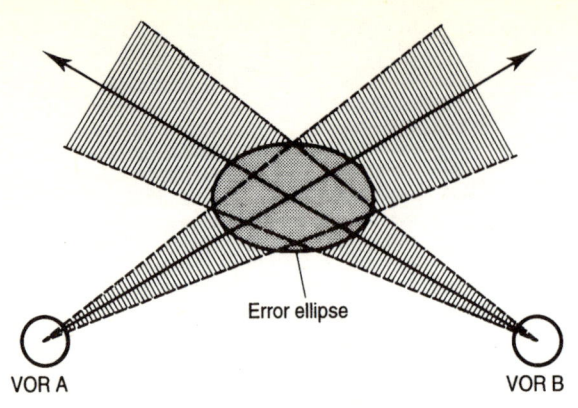

The Air Pilot's Guide to Satellite Positioning Systems

Fig 4 – Error Ellipse

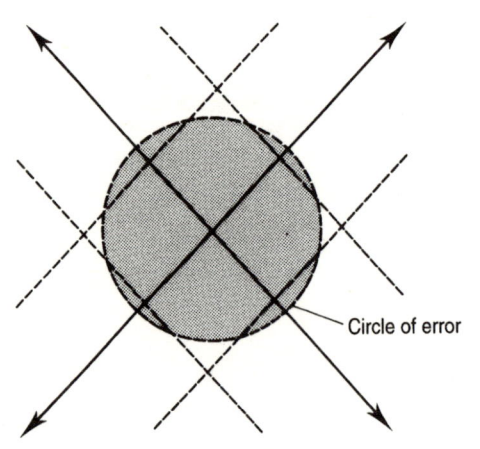

Fig 5 – Sources 90° Apart

cisely and there is always some error. If this error is shown as a shaded band as in Fig 4, then there is quite a large area that might contain the true position.

The other extreme is when the position lines cross at right angles, (Fig 5) causing the area of uncertainty to become a circle. It is now as small as possible.

Clearly, for best fix accuracy the two position lines should cross at right angles. This is always the case when range and bearing are obtained from a single point, although that has to be set against the range-dependent loss of precision of the bearing. More position-lines would not necessarily improve *fix* accuracy even assuming they were all measured to the same precision – it would still depend on their geometry. Where they would help would be if there were originally only two position lines and one was grossly in error. Just the two would give no indication even that an error existed but a third line would indicate something was wrong by producing a large triangle of inter-

Position-Finding for Navigation

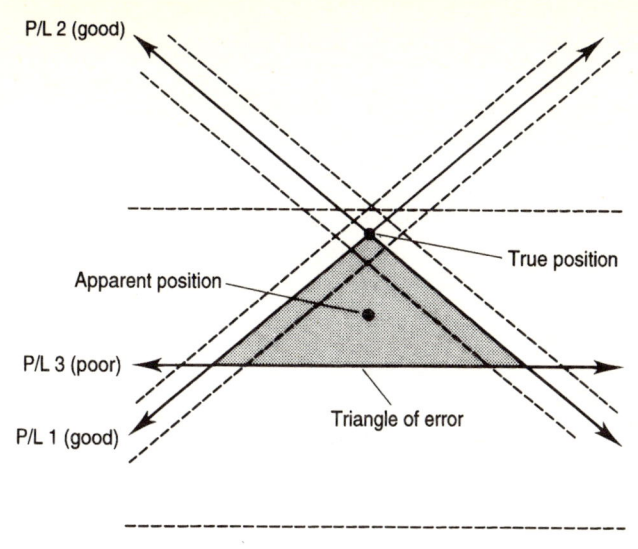

Fig 6 – Triangle of Errors

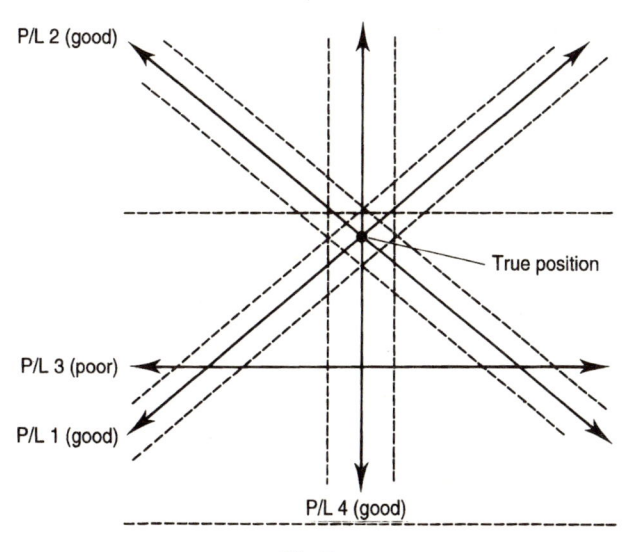

Fig 7

section as in Fig. 6, although it could not indicate which one was wrong. To do this a fourth line would be needed and more would confirm it, as in Fig 7.

Because GPS is capable of providing as many as ten ranges simultaneously, this basic idea can be used in three-dimensional form to determine whether any particular satellite is misbehaving (Receiver Autonomous Integrity Monitoring – RAIM – described in Chapter 8).

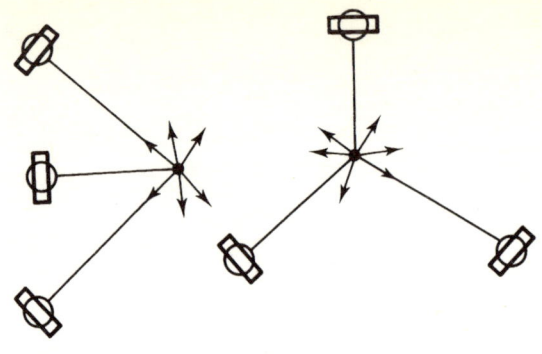

Fig 8

Statistically, if a number of position lines all intersect at the same point it indicates that the likely error of the fix is better than the precision of any one position line, each additional line reducing the fix error still further. However, there is a reducing return in that each time the number of lines is doubled, the fix error is reduced only by $1/\sqrt{2}$. Usually, 3 or 4 lines are enough to guard against gross blunders and give a useful increase in accuracy.

There is another effect due to geometry that is not often appreciated. Consider the situation where there are three position lines crossing each other at 60°. They could have been derived in two different ways – with all the transmitters off to one side at 60° separation or equi-spaced around the fix point at 120° separation (Fig 8).

If there are only negligible random errors there will not be much difference between the resultant fix errors wherever the transmitters are, but if there is a constant and similar offset error in all the ranges fix behaviour is quite different. When all the transmitters are to one side, the fix will be considerably displaced sideways and a navigator following the normal tendency to assume the middle of the triangle is the correct fix would end up with quite a large fix error, with only a small indication that anything was wrong (Fig 9).

But when the transmitters are spaced equally around him the centre of the triangle still indicates correct position (Fig 10).

This is also true if all ranges are too short, leading to the valuable fact that adding or subtracting the same amount from all ranges until the size of the triangle is minimised enables fixed errors *that are the same for all the lines of position* to be be taken out. Also note the difference in size of the error triangles – both figures are drawn to the same scale and assume the same fixed errors. With proper geometry there is much more sensitivity to small offsets and the correction is considerably more accurate.

Although this discussion has been confined to the two-dimensional case the same considerations also apply to the fully three-dimensional case of satellites. This is elaborated further in the RAIM section of Chapter 8, where the

Position-Finding for Navigation

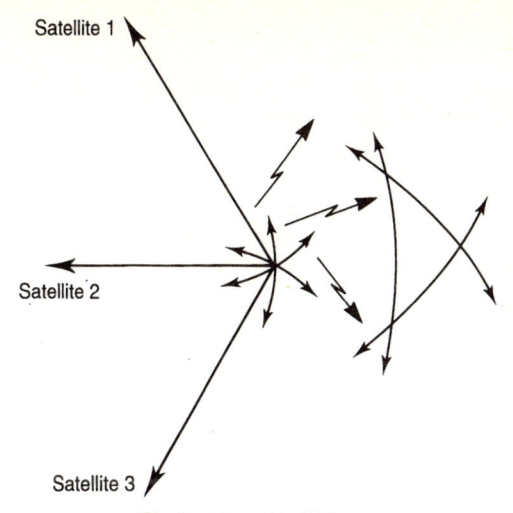

Fig 9 – One-sided Fix

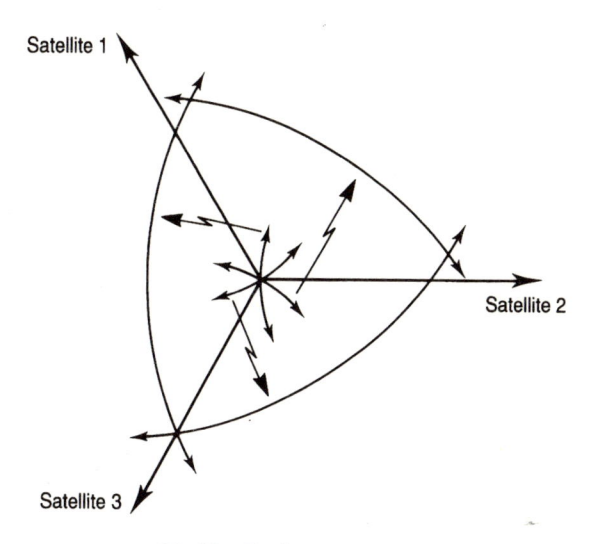

Fig 10 – Equi-spaced Fix

use of this principle to find faulty satellites is described. It is also used in some satnav receivers for local clock calibration.

There are therefore two constituents of navigation fix accuracy – the inherent precision with which a position line can be measured and the geometric effects that occur when combining two or more to provide a fix. Measurement precision of a single range tends to attract most attention but particularly in satellite systems it is much more likely to be geometric effects that determine system accuracy. Navigation systems designed to be used on their own without outside assistance ('stand-alone') always provide at least two simultaneous position lines, often three, sometimes more. Satellite positioning systems obey the same rules but with the added complications that:

> (a) they are inherently three-dimensional (i.e. they provide not lines but spheres of position), and
> (b) due to satellite motion their geometry is constantly changing even when the user is stationary, necessitating continuous re-determination of the best set of satellites.

4. The Derivation of Lines of Position

There has been little reference as to how the position lines used for getting a fix are actually to be obtained because as far as geometry is concerned it does not matter. In radio systems, they are obtained by the measurement of some physical attribute of radio waves with a geographical correlation, so, for instance, the early attempts to use the fact that they get weaker with distance. Measurements of their direction of arrival, or of the time lag between transmission and reception, are far more accurate.

4.1 Directional (angular) Measurements

One of the earliest and simplest ways of obtaining a position line was to take a bearing on a transmitter using a simple loop aerial. A more accurate way is to reverse the process – the transmitter sends a signal modulated in such a way as to permit a receiver to determine bearing just by 'listening' to it, as for instance in VOR. Both have the inherent problem that accuracy is dependent on distance from the transmitter. It doesn't matter much for homing because it gets smaller as the beacon is approached, but it is not very good for general navigation purposes such as RNAV. For this reason, bearing-only radio transmitters are usually located at some fixed point the navigator wishes to reach and are not intended as wide-area navigation aids. For RNAV, it is better to use range-measuring systems such as DME or PDME.

Satnav systems cannot be based on bearings because of the very high precision of measurement that would be needed due to the great distances involved. A GPS satellite is about 12,000 miles above the surface of the Earth and if a position line with an error of only 300 feet were required, then its bearing would have to measured to a precision of around .002 second of arc. This is quite impracticable with any sort of aerial that might be mounted on an aircraft or a satellite and for this reason bearing systems are not discussed further.

4.2 Range (distance) Measurements

Although not entirely safe to do so, it can be assumed for most radio-navigational purposes that the speed of travel of a radio wave is constant and known and can be used as an analogue of distance. The usual approximation is that radio waves travel at 300,000 kms per second, so to get a range measurement precision of 30 metres the timing system must be good to one-tenth of a microsecond (microsecond = one-millionth of a second). Although it sounds incredibly precise, it is not too much of a problem with modern technology. Then, all that has to be done is to measure the time a radio wave takes to get to the user and errors will arise only in the timing system itself. However, timing errors tend to be constant (at least over short periods) and are the same no matter what the actual distance involved may be. Therefore, the actual range to be measured is not important and errors will be the same whether it is a DME a mile or two away or a satellite 12,000 miles high. The method is admirably suited to satellite systems and is universally used by them. The engineering problem of how the actual measuring is to be done – elapsed time can only be measured if there is a start and a stop point – is discussed in Appendix 1.

5. Positioning Systems

A single position line cannot provide position on its own, so multiple transmitters working in concert must be used. The ways in which they can be combined are legion and have led to the development of many hundreds of different systems. Basically the requirement is that there should be sufficient transmitters available at any place or time to give the navigator at least two position-lines (in a two-dimensional system) having the correct geometrical properties. Because of this geometrical requirement, it is all too often the case that although radio reception is possible, a satisfactory fix cannot be obtained. To obtain the right geometry the intended area of use must be carefully defined and transmitters placed accordingly.

Exactly the same considerations apply to satnav systems, although it is complicated by the fact that the transmitters are moving and a satnav receiver must therefore continuously evaluate whether the best combination of satellites is being used.

6. Three-dimensional Effects

Most navigators are accustomed to using fixes plotted on a paper or electronic chart, and ignore such esoteric matters as how far they are from the centre of the Earth. This is quite permissible when using Earth-based radionav systems when transmitters are on, or nearly on, the same surface as the user. Bearings are not affected by height above the transmitter while slant range can easily be converted to plan range knowing height above sea-level. Only surveyors or others requiring great accuracy need to think about such details as the height of the transmitter above sea-level as well as aircraft height. Unfortunately, users of satellite systems cannot afford this luxury; their transmitters are great distances out in space and their 'height' is very important. The effects are described in Chapter 2, but here it is enough to note that 'ranges' normally thought of as curved lines must now be thought of as segments of spheres centred on the satellite. This has important effects on fix geometry, which becomes fundamentally three-dimensional.

Chapter 2

Basic Satnav Systems

The obvious difference between satellite and Earth-based radionav systems is that satnav transmitters are carried aboard satellites at great heights instead of being on the Earth's surface. Satellites are always in motion and the determination of transmitter position must be a continuous process instead of being done once and for all as with surface transmitters. All the effort involved in measuring the distance from a satellite very accurately will be wasted if it is not known where the satellite is. Also, as already seen, geometry plays a very important part in navigational fixing and satellites must be chosen to provide the correct relative positions for a good fix. Selection and measurement of orbits are therefore critical and choosing the right combination of orbits is the basic and most important decision made in designing a satnav system.

Following that, exactly how are ranges to be measured to a satellite thousands of miles out in space? A receiver has no way of knowing when it transmits so it can't simply do a sort of stop-watch measurement.

Even when it makes its measurement, will it be accurate enough to give a good fix?

And how will users know whether the system is working properly?

These are the topics to be discussed in this chapter.

1. Satellite Orbits

Most of us tend to think of satellites as circling the Earth, but this is liable to produce a false idea of what is actually happening. It is better to think of them as spinning round in their orbits like giant gyroscopes in space with the Earth at the centre of the gyroscope. This way, it is easier to appreciate the fact that, like gyroscopes, the plane of their orbit is more or less fixed with respect to the Sun and stars, (at least over a few weeks). Since the Earth revolves once every twenty-four hours, making the Sun appear to rise and set, satellites also appear to rise and set in the same way. This is most obvious with satellites in very low orbits which pass across the sky firstly low on one horizon; then going almost overhead; and finally low on the other horizon. Some of the low-orbiting weather satellites are kept exactly synchronised with the Sun so that they always pass over in the mornings or afternoons and provide weather pictures at the same time every day.

This spatial orbital stability is only relative and orbits change slowly over a period so that left to themselves satellites would slowly drift out of the desired orbit. They must be re-positioned from time to time and geostationary satellites need this to be done every few weeks so that they remain within the beam of the receiver aerials. They have small on-board engines to provide the necessary thrust and the fuel to do this is the major determining factor in their useful life because once expended the satellite's position cannot be maintained.

There are several reasons why orbits are not completely stable. The Earth is not perfectly spherical, so its gravitational attraction varies depending on where the satellite is; the Sun and the Moon exert gravitational attractions of their own which may cancel or reinforce each other; and there are several other minor effects. Even the radio transmissions from satellites produce small forces that over a long period push them off-orbit slightly. These effects are often quite small and for many communications satellites they can be ignored as long as they stay within the beam of their ground aerials or within a few hundred kilometres of their allotted positions. Navigation satellites are of course quite different; any positional error is reflected into the accuracy of the fix they produce. The target accuracy for knowing where the GPS satellites are is only six metres. Navigation satellites therefore need ground tracking networks that can measure their orbits continuously to a very high degree of accuracy.

1.1 Three-dimensional Satellite Positions
Satellite positions in space cannot be described in terms of the two-dimensional lat/lon and use three-dimensional systems. There are two major methods.

(a) The Keplerian System
If orbits are thought of as large gyroscopes in space, we can specify their diameter; how they lie relative to the stars; whether they are actually circular or not; and where the satellite is around the rim of the 'gyroscope' at any given time. The centre of the 'gyroscope' is the 'mass centre' of the Earth – not necessarily the geographical centre which depends on which geoid is being used. This is known as the Keplerian system and is the traditional method of describing satellite and planetary orbits. It has the advantages of being easy to use for forecasting orbits into the future and of giving a good intuitive 'feel' for what the satellite is doing, but involves some mathematical complexity in computing because it produces angular positions which have to be turned into the instantaneous x,y,z Earth-centred satellite positions which are what is really needed for measuring ranges.

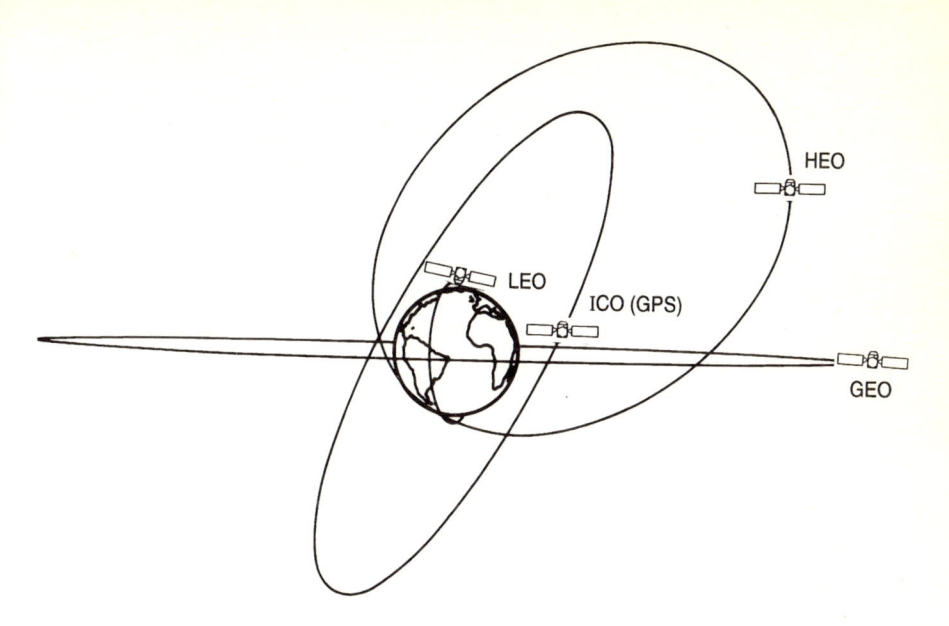

Fig 11 – Major Types of Orbit

(b) Geocentric Cartesian System.
Satellite position can also be specified relative to the mass centre of the Earth directly in a three-dimensional x,y,z co-ordinate system. Although this simplifies computation it gives no inherent information about orbital behaviour and can incur larger errors than the Keplerian method if used for future projections.

Generally, American systems have used Kepler's, while Russia has used x,y,z.

More information on these systems is given in Appendix 2.

1.2. Navigational Aspects of Different Types of Orbit
The major types of orbit are shown in Fig 11.

1.2.1 Low Earth Orbits (LEO) (up to 2,000 kms height)
These are the easiest orbits to achieve in terms of cost and size of launcher and being nearer the Earth stronger radio signals are possible for a given transmitter power. There is a lower altitude limit imposed by the Earth's upper atmosphere which produces excessive drag on the satellite and eventually causes the danger of the satellite prematurely entering the Earth's atmosphere and burning up. The upper limits of the atmosphere vary from day to day and day to night and satellites that are otherwise in 'safe' orbits are occasionally affected by it when a combination of factors make it 'bulge' upwards more

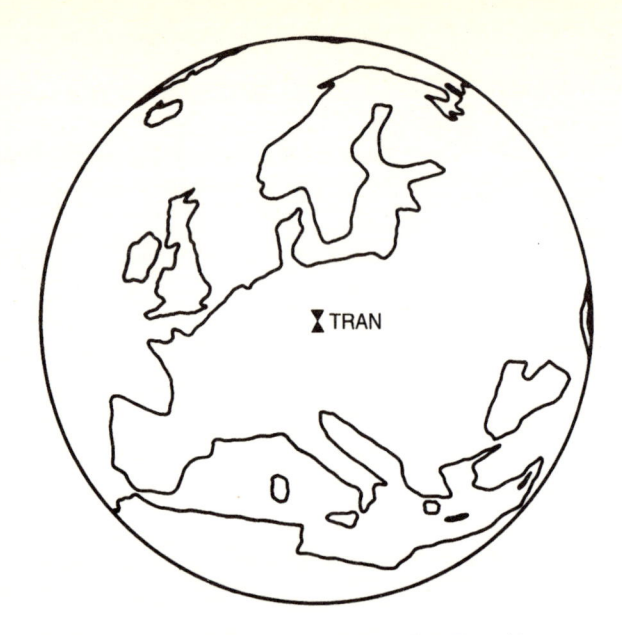

The Air Pilot's Guide to Satellite Positioning Systems

Fig 12 – Earth Coverage of a LEO (Transit)

than normal. These bulges cause orbital disturbances and added to local Earth gravitational variations make continuous orbital updating for navigational satellites in these orbits absolutely essential. The early satnav systems 'Transit' (see Appendix 3) and the Russian 'Tsicada' used such orbits mainly because they were all that could be achieved with the launch vehicles available at the time they were designed. The later 'Transit' satellites used a system of active on-board orbital compensation (the 'Discos' system) in order to try and maintain more accurately predictable orbits.

Advantages for navigation:
    a. Relatively low-cost and simple spaceborne and user equipment.
    b. Radiation effects are low.
    c. Low altitude provides rapidly varying Doppler-shift which can be used for navigational purposes.
    d. Being close the the Earth's surface, low-power radio transmitters can provide adequate signal strength for simple receivers.

Disadvantages:
    a. The orbital period of around 105 minutes means that the satellite is only in sight for about 15 minutes once every 105 minutes, and continuous world-wide navigation cannot be achieved without using a very large number of satellites (about 86).

Basic Satnav Systems

b. Relatively large and rapid variations of orbital position that are difficult to predict.
c. One satellite's transmissions cover only a relatively small area.

1.2.2 Intermediate Circular Orbits (ICO) (5,000 – 20,000 kms)

These heights provide considerable scope for variation in orbital period which may be from six to twelve hours and one satellite can cover a large area and remain visible from the ground over a long period. The area of coverage is not much less than that of a geostationary satellite – compare Figs 13 and 14. Non-uniform Earth gravitational effects are much reduced as are residual atmospheric drag effects.

Advantages:
    a. In view for reasonably long periods requiring less satellite-swapping.
    b. Stable orbits allowing more accurate long-term orbital predictions.
    c. Large area of coverage from any one satellite – needs far fewer for continuous coverage than LEO (20–30).

Disadvantages:
    a. Slower Doppler-shift.
    b. More expensive in launcher costs.
    c. Some radiation effects.

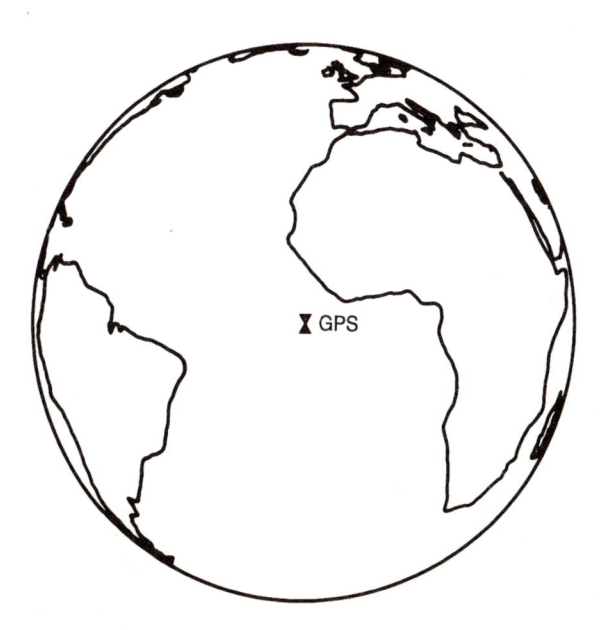

Fig 13 – Earth Coverage from an ICO (GPS)

The Air Pilot's Guide to Satellite Positioning Systems

1.2.3 Geostationary Earth Orbit (GEO) (36,000 kms)

Although much used for communications purposes, this orbit is not particularly useful because of the geometric problems inherent in single-plane systems (described in more detail later). A world coverage three-dimensional satnav system using only geostationary satellites would be impossible, but the addition of signals from geostationary satellites to a satnav system based on other types of orbit can be of considerable value.

Advantages:
    a. For a two-dimensional system covering specific areas below the Poles requires the smallest number of satellites (3–4).
    b. Comparatively stable orbits.
    c. Navigation payloads can be piggy-backed on communications satellites to cut costs.

Disadvantages:
    a. Geometric problems prohibit fixing around Equator.
    b. Cannot provide height.
    c. Cannot cover Poles.
    d. Satellite and launcher costs high.

1.2.4 Highly Elliptical Orbits (HEO)

These orbits have a pronounced elliptical shape (large eccentricities). Their main advantage is that at apogee (their most distant point) they appear from

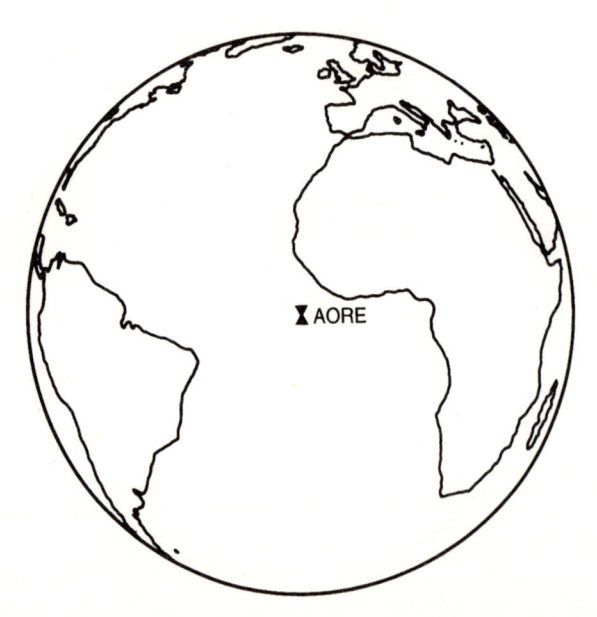

Fig 14 – Earth Coverage from a GEO (Inmarsat Atlantic E Satellite)

the Earth to stay almost stationary for several hours, and this point can be arranged to be overhead areas far North or South that cannot be covered by GEOs. This property makes them useful for delivering signals to vehicles operating in built-up areas in northern or southern cities where satellites below about 30°–40° elevation cannot be seen due to blockage by buildings. Like GEOs, HEO satellites in a single plane do not provide sufficiently good geometry for navigation and satellites in other planes or types of orbits must be added.

Advantages:
    a. Can provide good signals to users in high latitudes.
    b. And to mobile users in lower latitudes in 'canyon' areas.

Disadvantages:
    a. Poor geometrical properties.
    b. Expensive.
    c. Large number of satellites needed even for a small area.
    d. Very high Doppler-shift can cause receiver design problems.
    e. Only part of orbit usable.

(The view from an HEO varies between that of a LEO – Fig 12 – and that of a GEO – Fig 14. The extra apogee height of an HEO over a GEO makes very little difference.)

1.3 Summary

A system must incorporate several different planes of orbit to overcome geometric problems. Thus, GEO satellites on their own are unsatisfactory and can only be used in combination with satellites in other planes, while the control of an HEO system would be very complicated. Circular-orbit satellites are preferred for the main system and the choice is between LEO and ICO satellites. On the whole ICOs are preferred since they need fewer satellites and their orbits are more predictable and stable. Both GPS and GLONASS have adopted ICO 12-hour orbital patterns and the projected INMARSAT augmentation satellites may use 6-hour orbits.

2. How Many Satellites?

Obviously a system should be designed to use as few as possible consistent with its purpose. If it needs to cover only a single continent and can dispense with world-wide coverage then it may be able to use relatively few, but the number grows rapidly for full coverage. The first thing to be determined is how many are needed in sight at any one time to give a navigational fix – the geometric factor – then the system can be designed to keep this number in sight everywhere at all times in the desired area of coverage.

2.1 One Satellite

A single satellite could only provide continuous fixing if both range and bearing could be measured simultaneously, but it is impossible to measure bearing with enough accuracy. A single range is, in space, a sphere of position that results in a circle of position where it intersects the Earth's surface (unless the satellite is vertically overhead, when it becomes a point). This is not very useful on its own unless combined with position lines derived in other ways and consequently a single satellite cannot provide a navigational fixing service. It may be thought that Transit, in which a fix is obtained from a single satellite, is an exception, but not so. Range cannot be measured instantly and is established by taking a series of Doppler measurements, which take a few minutes and do not fulfil the requirement for an instantaneous fix whenever needed (for a full description of how Transit works, see Appendix 3).

There are other problems, particularly when applied to aircraft. Aircraft are not stationary, so the Doppler-shift caused by their own motion has to be calculated from accurate knowledge of their speed and heading. Also, Transit uses low-Earth orbiting satellites which come round only once every two hours, so there is a long time between fixes. For these reasons, Transit has never been used by aircraft operationally. However, the Doppler effect can be, and is, used in newer satnav systems like GPS, for deriving velocity rather than position.

These limitations are unacceptable for a navigational service that is to provide instantaneous, continuous and independent fixing.

2.2 Two Satellites

Provided they are well separated they will give two spheres of range which will intersect around a circle in space. If an observer on the Earth's surface knows his geocentric height then he has in effect a third 'sphere' of position (the geoidal surface of the Earth, in quotes because it is not quite a sphere) which will intersect the satellite-originated circle of position at two points that are usually so far apart there is no problem selecting the correct one (Fig 15).

Any inaccuracy in calculating geocentric height will of course reflect into position in just the same way as range errors would do, but otherwise this is a feasible system and is used in two-GEO satellite systems such as Geostar, Locstar, Omnitracs, and so on. The problems lie in obtaining geocentric height for a position that is not initially known, and in the actual measurement of range to satellite, which usually involves the user transmitting (see later).

2.3 Three Satellites

These will give three spheres of range providing an unambiguous fix (Fig 16) without needing any knowledge of geocentric height. Actually, they also produce a second possible position, but it is well out in space and obviously wrong for near-Earth users.

Basic Satnav Systems

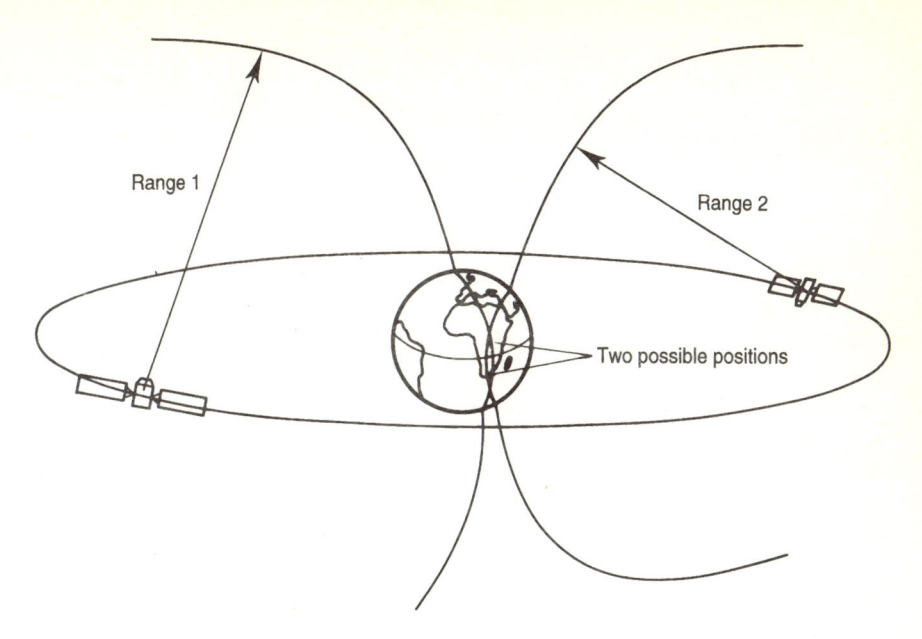

Fig 15 – Two-satellite Fixing

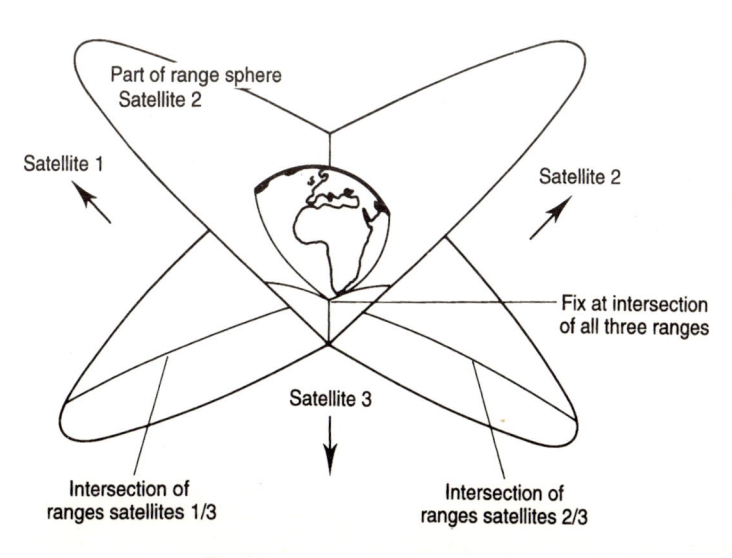

Fig 16 – Three-satellite Fixing

The Air Pilot's Guide to Satellite Positioning Systems

To get an absolutely unambiguous 3D fix four satellites are required.

This is the irreducible minimum number of satellites needed for an independent fix that does not rely on anything else. The fix is truly three-dimensional in space and is related only to the three satellites and the mass-centre of the Earth. Because of this there are some problems relating it to position on the surface of the spheroid and even more relating it to grids printed on maps, like lat/lon. They are discussed later in this chapter.

The angles of intersection of the three range spheres vary considerably depending on the relative positions of the satellites, so the actual accuracy of the fix will vary. For a specified fix accuracy they can only provide the right geometry for a short time and it is a matter for the system designers to decide how far geometry can be allowed to decline before a switch must be made to different satellites. This requires additional satellites to be put up in such positions that they will move into the right positions as the old ones drop out, increasing the required number of satellites quite considerably over the number needed simply to ensure there are three or four up over the horizon at any time.

Also, this minimum number of three is arrived at only from geometrical considerations. Other factors, such as actually obtaining ranges, may necessitate more.

3. Measuring Distance to Satellites

Having decided the minimum number of satellites we now turn to a consideration of how exactly the necessary lines of position will be obtained. Since taking bearings on a satellite is impracticable for a mobile vehicle, the only way of obtaining them is to measure range, and that in turn means measuring elapsed time very accurately since time is the only analogue of range we can use. To do so means that the timing system must have a start and a stop point and although the stop point can easily enough be obtained as the time when the satellite signal reaches the receiver, how is it to be correlated to the start point, the time of emission by the satellite? And if there are several different start points, one for each satellite, how are they to be accounted for?

In Fig 17:

> SD = Satellite timing clock offset from zero.
> TD = Satellite signal transit time to receiver.
> RD = Receiver clock offset from zero.
> RT = Measured signal transit time at receiver.

The fundamental problem is shown in Figs 17 and 18. In Fig 17, the system reference time is denoted as 'zero time' and is simply some arbitrary point of absolute time adopted as a reference for timing purposes. The satellite and receiver both have their own internal timing systems which are assumed for

Basic Satnav Systems

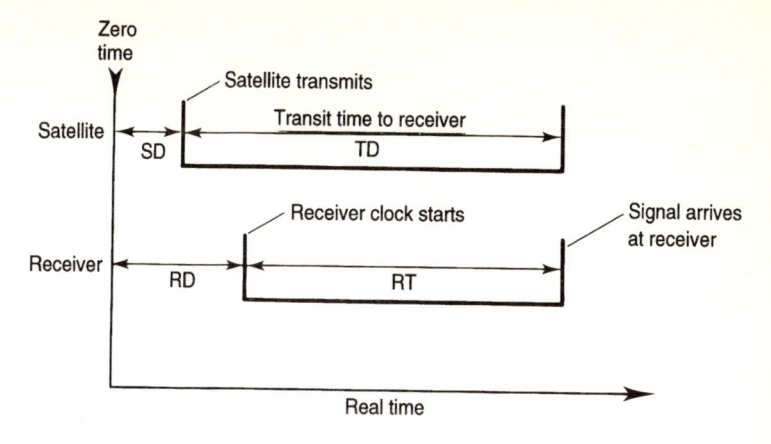

Fig 17 – Timing (1)

this purpose to be completely driftless and also to be precisely accurate and the same as each other as far as elapsed time is concerned. They are not aligned to absolute or system zero time, and this is why the satellite transmission and the receiver clock reference are shown occurring at different times relative to system zero.

What we need to know is time 'TD', which is 'RT' plus 'RD' minus 'SD', but we know neither 'SD' nor 'RD'.

Even if we could measure 'TD' we would still not have a fix if we could only do it for one satellite, and as just seen a minimum of three is essential.

Rather than complicate the drawing too much, in Fig 18 only two satellites are shown, but the principle can be extended to any number. Both satellites are assumed to be on the same basic time-base but their times of transmission are not synchronised either with each other or with the receiver.
In Fig 18:

> SD1 and 2 = Satellites 1 and 2 timing offsets from zero.
> TD1 and 2 = Satellites 1 and 2 signal transit times to receiver.
> RD = Receiver clock offset from zero.
> RT1 and 2 = Measured signal transit times at receiver.
> D = Difference in times of arrival.

In this multiple satellite system there are two options:

> 1. If both TD1 and TD2 were measured there would be two lines of position and therefore a fix (provided we also know geoidal height). To do this requires us to know SD1, SD2 and RD in advance.
> 2. D could be measured to provide the differential time of arrival between the two satellites by simply subtracting RT1 from RT2. This would only provide one line of position but we would not need to

The Air Pilot's Guide to Satellite Positioning Systems

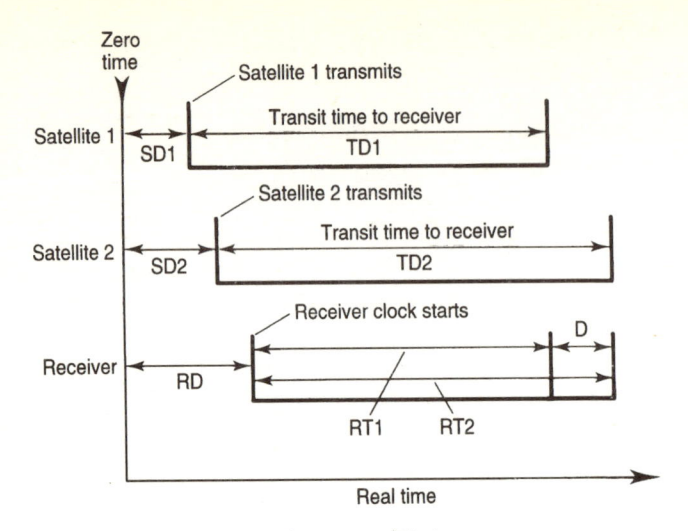

Fig 18 – System Timing

know any other timing data. SD1, SD2 and RD would be immaterial. To provide a two-position line fix another satellite would be needed, and for an independent 3D fix yet another one, a total of four.

3.1 Round-trip Ranging

The easiest way (conceptually, at least) to solve the ranging problem would be to transmit a signal from the receiver site that would be either reflected or transponded off the satellite back to the receiver. The receiver would use the same time-base as the transmitter and therefore would have its start point well defined; the stop point being simply when it got its signal back again. Referring back to Fig 17, there would be no 'RD', 'SD' would be the same as 'TD', and 'RT' would be equal to 'SD' plus 'TD'. Therefore, divide 'RT' by two and we have range to satellite.

Several satellites could be handled in sequence if the aerials could be turned around to align on each satellite quickly enough but it is partly this which makes direct round-trip ranging impractical for a satnav system although it is often used for determining satellite orbits. No less importantly, it would require a high-power transmitter and high-gain aerial and would mean that every user would have to transmit. The likely result would be that they would simply jam each other quite apart from the cost, complexity, weight and lack of portability of the equipment. High-gain dish aerials as used for portable satellite communications systems would be unusable because they would need to be continuously and very rapidly (maybe ten times per second) pointed in succession at up to eight or ten satellites in different parts of the sky. In turn, this would mean the receiver having to acquire and lock satellites

Basic Satnav Systems

very rapidly indeed. And the less said about aerodynamic effects on a small aircraft the better! A much smaller omni-directional aerial would not need to be mechanically re-directed in this way, but it would not have the radio power-gain of the highly directional dish aerial. Without this gain, to obtain the necessary radio power to get to a satellite perhaps 20,000 kms away would need quite a powerful transmitter which would be impractical in most aircraft.

In any case the system would become saturated by all the hundreds of thousands of users who might want to use it at the same time, and military users would not want to transmit because it would give them away. Nonetheless, some civil satellite positioning systems provide for the user to transmit a ranging signal, as shown in Fig 19.

In these systems a central control station transmits a radio signal carrying a timing mark which is re-broadcast by a satellite and picked up by a user vehicle which transmits back again to the master via the satellite using the incoming signal for timing its reply. The master station receives the reply and measures the round-trip time delay. All the timing and computation is done at the master station, and the satellite does nothing except change radio frequency and re-broadcast the uplinked signal. The vehicle unit can also be quite simple, having no need for accurate timing, the only requirement being that its time delay in re-broadcasting is kept constant. This may seem to contradict the earlier statement that in this sort of system vehicle units would need to transmit at high power, but the big difference is that because these systems normally cover only a rather small area the satellite aerials themselves are highly directional and have considerable gain that makes up for the vehicle's low-power transmission. One range established in this way does not provide a fix; another one is needed. Practical systems of this type are described in Chapter 7.

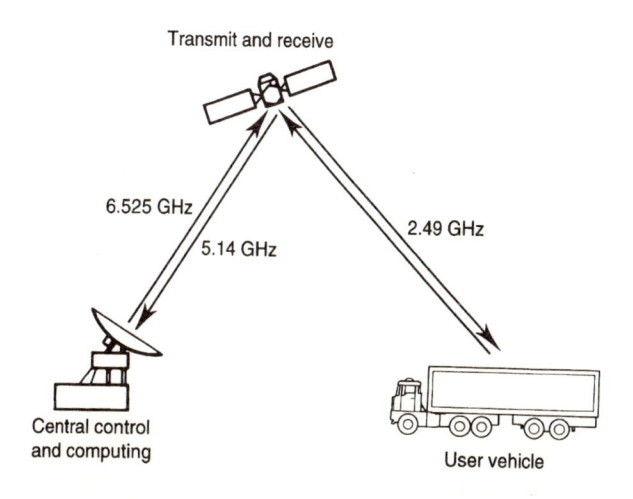

Fig 19

True wide-area satnav systems cannot afford to limit their coverage in this way nor can they risk saturation from (possibly) millions of users interrogating them so other methods of ranging must be found.

3.2 Absolute Timing Ranging

Range might be measured directly in the form of a time delay if the whole satnav system, satellites *and receivers* alike, could be directly and very accurately locked to Greenwich or some other standard of absolute time so that, for instance, all their clocks were at 10:00:00.000 hours (within a few million-millionths of a second) precisely at the same time. Then the time delay over the transmission path could be measured simply by noting the absolute Greenwich time at which the satellite signal was received.

For instance:
Satellite transmits signal	10:00:00.000 hours.
Arrives at the user at	10:00:00.001 hours.
	(1 millisecond delay).
Satellite range	300,000/1,000 kms.
	= 300 kms.

If receiver clock is always within 1 microsecond of satellite clock then accuracy would be 300,000/1,000,000 = 300 metres.

In Fig 17, 'SD' and 'RD' would disappear, making 'RT' equal to 'TD', which is what we want. This method could cope with any number of satellites and any number of receivers provided some method of identifying which satellite was which was incorporated. A nice idea but unfortunately quite impractical because there is no way of getting all the satellites and receivers directly on to a common absolute time-base. Even if some way of getting the satellites on to Greenwich time could be devised, each receiver would need correcting to Greenwich every time it was switched on, its switch-on time obviously being a random event. But since the receiver would not know its position it could not correct for the transit time delay in reception of the Greenwich time standard by radio and there is no other way of doing it.

The only remotely practical way of adapting this principle would be if the receiver had a clock that was extremely stable (like an atomic clock) and could be synchronised with GMT in advance before it was needed for navigation. For instance, if it were located at some accurately known point it would then know the absolute time delay and could correct for it. After that it could move away and continue ranging. Operation in this mode would be rather like setting up an aircraft's inertial system while on the stand at an airport and hoping it does not drift off too much in the subsequent flight. In fact, accuracy would be rather similar and it would offer no real advantage over the existing INS.

This might be described as 'external time-base' ranging, as against 'internal time-base' as used by Decca, Loran and all differencing systems.

3.3. Differential Ranging

If direct ranging as described in 3.1 and 3.2 is impossible then some other common timebase must be provided. Methods of differencing transmitters were described in Chapter 1 sections 4 and 5 and are equally applicable to satellites. Fig 18 shows how two satellites can be differenced against each other in the receiver to provide 'D'. A third satellite would be needed to give two range differences, requiring knowledge of geoidal height to get a fix, or a fourth satellite to get three and a stand-alone fix. When differencing is used the range differences are hyperboloids, but this does not affect the fix requirement. The only requirement is that SD1 and SD2 stay absolutely constant relative to each other, in other words the satellite transmissions are all locked to each other. This could be done either by using such excellent clocks that they could do so without outside assistance (a real possibility for the future) or directly by inter-satellite links. This technique has not yet been used partly because of technical problems arranging inter-satellite timing systems that will work with a large number of satellites, but the GPS Block IIR satellites will incorporate such a system. When satellites lock to each other's transmissions in this way there is really no need for transmissions from the ground, except at long intervals, which is desirable from a military viewpoint. It also eliminates the need for receiver clock calibration and makes for simpler receivers.

3.4. Pseudo-ranging

The method adopted by GPS is 'pseudo-ranging', so called because the ranging measurement made in GPS receivers is composed of many other factors than just the range of the satellite from the user. But as far as users are concerned, pseudo-ranging has the same results as if direct ranging had been employed. First, all satellites are referenced to a common time-frame by calibrating their timing against a central standard. A monitoring station tracks their signals and determines how far off the system's time standard (which might indeed be an international standard like Greenwich) they are. It can do this because it knows both its own position and that of the satellite with great accuracy and can allow for the time-lapse in reception. Not being in motion it can also average over some little time and further increase accuracy. Then, instead of correcting satellites directly it transmits corrections to users via the satellites. The end result is the same as if all satellites were in fact operating on the same time standard. In the terms of Figs. 17, 'SD' plus 'TD' is measured at the monitor site and 'TD' subtracted to give 'SD', which is then sent to the user as a digital correction.

This puts all the satellites on the same time-base but the receiver is still outside the timing loop, so the next step is calibrate 'RD' – the receiver timing (clock) offset and to see how this is done we must delve into a little mathematics.

Assuming all positions are specified in a common three-dimensional (x,y,z) frame centred at the Earth's geocentre, and using the nomenclature of Figs 17 and 18, then:

If satellite position is X_s, Y_s, Z_s (units immaterial)
and user position is X_u, Y_u, Z_u
The physical distance R_p between them is, by simple geometry:
$$R_p = SQR[(X_s-X_u)^2+(Y_s-Y_u)^2+(Z_s-Z_u)^2]$$
and the measured range R_m is:
$$R_m = C/(RT - [RD - SD])$$

Where:
- C = Speed of light
- RT = Measured delay time between receiver reference clock and receipt of satellite signal
- RD = Unknown time offset between receiver clock and system clock.

The user position, X_u, Y_u, Z_u, is what we want to know, but (RD–SD) is also unknown. This makes four unknowns for which there is only one equation, so we need three more similar equations, which can be obtained by measuring T_d to three more satellites.

$$(X_1-X_u)^2+(Y_1-Y_u)^2+(Z_1-Z_u)^2 = C/(RT - [RD - SD])$$
$$(X_2-X_u)^2+(Y_2-Y_u)^2+(Z_2-Z_u)^2 = C/(RT - [RD - SD])$$
$$(X_3-X_u)^2+(Y_3-Y_u)^2+(Z_3-Z_u)^2 = C/(RT - [RD - SD])$$
$$(X_4-X_u)^2+(Y_4-Y_u)^2+(Z_4-Z_u)^2 = C/(RT - [RD - SD])$$

Where:
$X_{(n)}, Y_{(n)}, Z_{(n)}$ = Satellite co-ordinates for satellites 1,2,3,4.

This enables a solution for (RD–SD) but since 'SD' for each satellite is part of the downlinked message it can be taken out and the result is 'RD' alone. We then have (in Fig 17) 'RD', 'SD', and 'RT' and can therefore obtain 'TD', the desired result.

The receiver clock has only to remain stable for the few milliseconds while measurements are being made. If very high accuracies are not being sought then the small drift in the clock that might occur in this time is ignored, but for higher accuracies it is calibrated and a correction applied.

On the other hand, the satellite clock must be very stable because it has to maintain timing stability between the periods when it is calibrated by the ground monitor.

3.5 Pseudo-ranging Two-dimensional Systems

The method just described requires that the satellites carry extremely stable clocks and means that the satellites must be specially designed for the job. If it is possible to upload continuously from the ground, then this requirement

Basic Satnav Systems

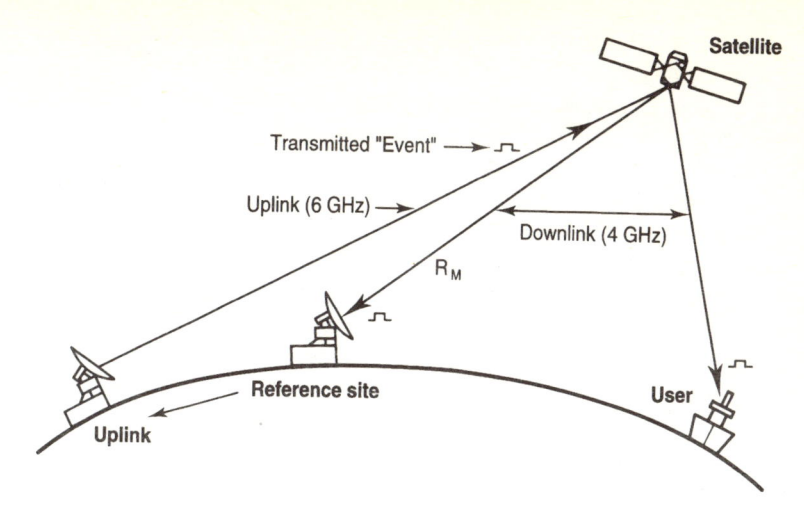

Fig 20 – PR 2D Fixing

in the satellite is taken away and much simpler navigation payloads can be used, but the satellites must always be in view of upload stations. For fixed upload stations this is only true if the satellites are geostationary but as already seen, the geometric properties of GEOs are not all that favourable for independent fixing because they are all in a single plane. If however the spheroid can be used to assist three-dimensional fixing then the geometric situation improves considerably. For some applications, knowledge of geocentric height is not a particular problem and a two-dimensional fix is perfectly acceptable. For instance, in marine/offshore surveying, it is often quite accurately known for other reasons and it is easy to use it for navigational purposes. A system of this type is shown in Fig 20.

As in direct ranging systems, a master station generates a timing signal and continuously uplinks it into the satellite(s) which then re-transmit it to be received by users and by a number of monitor stations whose positions are precisely known. These stations send time-of-arrival data back to the master station and thereby complete a timing loop. Really, it is rather similar to the round-trip ranging described in 3.1 apart from the return signal going through intermediate 'repeater' stations (the monitors). The master thus has timing data on all satellites in the system and by comparing actual TOAs with those it would expect theoretically, it can correct any transmission-path errors caused by refraction and local uncalibrated delays. Knowing also the orbital positions of the satellites, the master station can calculate the time delay its signals undergo on their way up to the satellites. This information is transmitted to the users as part of the uplinked signal permitting them, in effect, to subtract it from the overall TOA and make the system behave as if the timing originated at the satellite.

So far there is little difference between this and direct-ranging, but the major difference occurs at the user in that he is not required to transmit; he simply receives the satellite signals passively. This takes him out of the basic system timing loop which, as in GPS, creates a need to eliminate timing differences between his receiver and the rest of the system.

This is done in much the same way as described for GPS in 3.3 above. If the timing offsets of each satellite are known then a two-dimensional position necessitates only three range equations which can be derived from three satellites and will provide two spheres of position.

The main advantage of this type of system lies in the comparative simplicity of its satellite arrangements. The satellites themselves do nothing except relay the uplinked signal and consequently can be very simple – existing transponders on communications satellites are often used. The only extra burden placed on them by their navigational use is that their positions must be much more accurately known than is necessary for their communications job.

3.6. GPS and GLONASS

Pseudo-ranging systems continuously uplinked from the ground have the advantages of allowing the satellite segment to be very simple and keeping all the computing and timing systems on the ground where they can be easily accessed. However, for military purposes it is unacceptable to continuously uplink, being too easily jammed or spoofed, so the satellites must be designed to have some degree of autonomy, that is, operate for some time without ground uplink or correction, and pseudo-ranging is really the only option unless they can be linked together.

The two problems that arise in autonomous operation are that their orbital positions drift away from those predicted, and their clocks drift off calibration. How long they can operate without an uplink is a function of how well these two problems are corrected.

Intermediate circular orbiters at heights of 20,000kms as used by these systems are not nearly as prone to short-term orbital disturbances as are low Earth orbiters and their orbits can be projected reasonably well into the near future. As long-term experience is gained with these systems these projections are steadily becoming more accurate and it is thought that the current GPS system could operate without upload for as long as a week. The real answer is to enable the satellites to perform their own orbital determinations and in the GPS Block IIR and F satellites, provision is being made for inter-satellite tracking which it is forecast will enable them to operate for up to six months without uploads.

Clock drifts are not as easy to compensate. To take some figures, if the pseudo-ranging error due to clock drift is not to exceed, say, three metres, then the on-board clock must not drift more than ten nanoseconds between uploads. If the on-board clock is a rubidium atomic standard, its mean drift will probably be in the region of 1 part in 10^{-13}/sec so the maximum period between uploads would be about thirty hours. This assumes that the drift is unpredictable but actually once the satellite has been in orbit for some time the behaviour of its timing standards often become quite predictable and steady drifts can be signalled through the standard satellite message.

Once again, the answer lies in inter-satellite ranging and time-locking. For the satellites to keep themselves all on the same time-base requires either that they know the physical distance between themselves very accurately or that they can all lock to some non-GPS time-base simultaneously. Methods of doing this are being investigated.

4. Dilution of Precision (DOP)

If there were only the minimum number of satellites in orbit required for an independent 3-D fix (4) then ideal geometry would occur only momentarily as the satellites happened to come into the right positions and at all other times they would be at less favourable positions. The solution is to have many more satellites placed into orbits in such a way that as satellites drift out of the optimum positions another drifts in. The problem has always existed in all positioning systems but in non-satellite systems it was a two-dimensional problem, exemplified by the crossing angle of the hyperbolae in Decca or Loran becoming increasingly acute as range increases. Satellites operate in three dimensions and the problem is more difficult. The spherical segments of ranges from four satellites make a tetrahedron in space, the size of which depends both on the accuracy of measurement and on the geometry involved. Their combined effects are summarised as a non-dimensional number called Geometric Dilution of Precision (GDOP), while if only three-dimensional geometry is considered, the number becomes Position Dilution of Precision (PDOP). Practical navigation is usually more concerned with accuracy in the vertical or horizontal planes than three- or four-dimensionally so it is the projection of the tetrahedron (or polyhedron if there are more than four satellites) on to a flat plane that we want. When it is projected on to a plane tangential to the surface of the Earth it is called Horizontal Dilution of Precision (HDOP), while if onto a plane in the vertical it is Vertical Dilution of Precision (VDOP).

Although this terminology first appeared in connection with GPS it is now being used for other satnav systems, but it was never necessary for ground-based systems because they are not 3-D systems and only HDOP was really relevant. Another new GPS technique is a method whereby the receiver itself can work out how accurate its fix might be. If it works out whatever DOP is appropriate it only needs an estimate of the actual ranging errors being encountered on the satellites in use to do this. In GPS these are transmitted as the User Equivalent Range Error (UERE) number in the satellite message

Table 1 – Dilution of Precision.

Acronym		Dilution of precision due to:
GDOP	(Geometric)	3-Dimensional geometry plus timing errors
PDOP	(Position)	3-Dimensional geometry only
HDOP	(Horizontal)	Horizontal geometry
VDOP	(Vertical)	Vertical geometry
TDOP	(Time)	Timing errors only

(which allows for Selective Availability errors). So, if the HDOP for a given set of satellites is, say, 2, and the UERE is 32, the accuracy of a fix in the horizontal plane would be unlikely to be much better than 32*2 = 64 metres. (This somewhat simplifies the actual mathematics but gives the general idea.) The various DOP figures being experienced at any time are provided on many receiver read-outs.

There is another geometric factor to be considered in satellite systems (although it is really only a special case of poor PDOP). When satellites are orbiting in only one plane their 'spheres of position' intersect only at very acute angles for users in the same plane as the satellites, and any geocentric height error produces abnormally large errors (Fig 21).

It becomes so bad in areas directly underneath the satellites that they are unusable there. The problem can be solved by the addition of more satellites not in the same plane of orbit, which is why GPS and GLONASS use several different orbital planes. Users near the Equator attempting to use a system based entirely on geosynchronous satellites would encounter this problem, whether or not they used geoidal height as an input. The 'Starfix' system, for example, is unusable near the Equator.

5. Determination of Satellite Position

Low Earth orbiting satellites can be directly launched into orbit but for higher satellites it is more economical and precise to place them first into a transfer orbit before final ejection into their final orbit. The procedure is that the initial launch puts them into a low-Earth circular orbit after which the main launch vehicle falls away. When the orbital characteristics are established, the space vehicle's own much smaller motor is fired to put the vehicle into a highly elliptical orbit having its apogee at the desired orbital height. Once this has stabilised, a final firing of the motor while at apogee circularises the orbit at the apogee height. This is known as the Hohmann transfer orbit procedure.

This initial launch data is used to determine the approximate Kepler parameters for a first orbit. Although not nearly accurate enough for navigation, it enables ground tracking stations to find the satellite and perform further track-

Basic Satnav Systems

Two geosynchronous satellite fixing

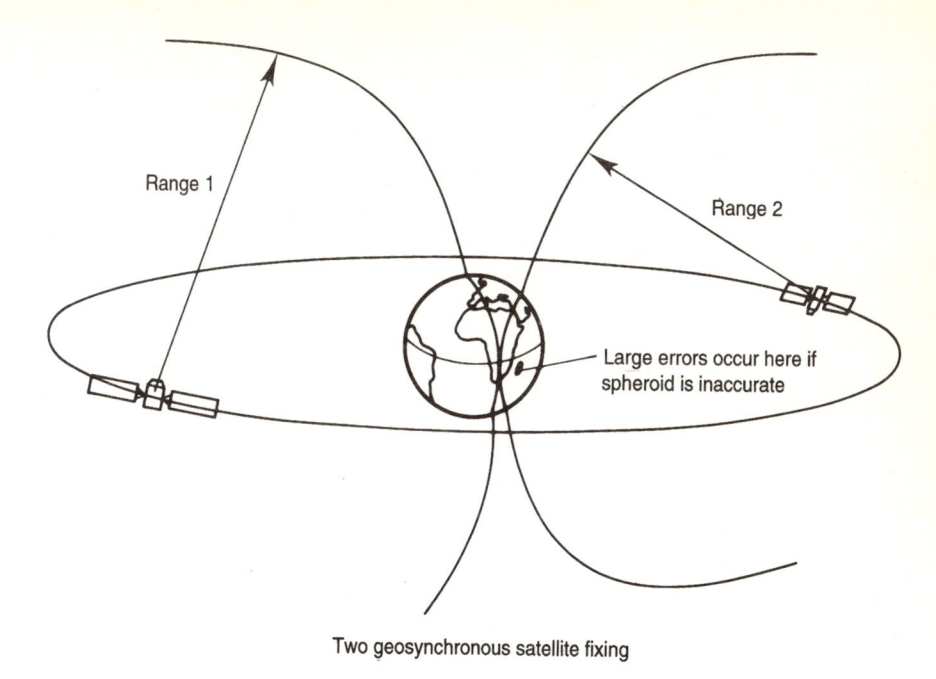

Fig 21 – Single-Plane Accuracy

ing. In order to establish satellite position more precisely a reverse process to navigation is used. Ground stations whose position is known, track each satellite and measure ranges over a period of time. These ranges are then used in reverse triangulation to determine the position of the satellite. Because these stations are stationary, considerably more data-filtering can be applied than in a rapidly moving mobile receiver and more accurate time references and receivers can be used. A succession of satellite positions enables refinement of the initially adopted Keplerian parameters and derivation of an accurate orbit. As time goes by increasingly accurate orbits are derived by using earlier estimates and refining them. When the refinement of orbital information – the ephemeris data – has proceeded sufficiently to be usable for navigation, and it has stabilised, it is uploaded into the satellite and re-transmitted to the user receiver. The entire process from launch to usability for navigation may take a month or so.

Great care is taken at the monitor stations to correct for all possible perturbations in the measured data. Ionospheric and atmospheric refraction, relativistic time shifts, lunar and solar gravity effects, and polar wander are all accounted for. After the monitor station data is received at the Master Control Station, further correction and filtering may be applied, using data from several monitors simultaneously, before it is used for final orbit determination. In GPS, this process results in satellite position always being known to better than six metres.

From these measurements and knowledge of how position varied on previous orbits it is possible to predict with high accuracy where they will be for some time ahead. This is uploaded into the satellites' memory and re-transmitted as part of its message together with other data. When transmitted this way it is known as the 'broadcast ephemeris' and is intended for those users who need real-time fixes. It is a *forecast* of satellite position and as such is subject to small, but usually navigationally insignificant, errors.

Many surveyors and those using the system for timing purposes are more interested in high accuracy and do not need real-time results. For them the broadcast ephemeris may not be accurate enough but they can afford to wait a few days for more accurate data to become available. This is known as the 'Precise Ephemeris' and is the result of actual tracking data rather than a prediction. It also incorporates data from many more tracking stations and can allow for the real ionospheric conditions experienced at the time.

Geosynchronous communications satellites with navigation payloads need additional tracking over that installed for their communication role. They are rarely exactly on the Equator and inclinations of a degree or two are not unusual. This together with out-of-circular eccentricities of .0003 or so produce variations of several hundred kilometres in their positions. As an example of how their positions may vary, Table 2 shows how the elevation and azimuth of Inmarsat satellites vary over a day as seen from London.

Navigational systems using GEOs (Starfix, Locstar, etc) use round-trip ranging and not pseudo-ranging as in GPS, so the effect of errors in satellite position on the user are not nearly as serious as they would be in GPS, but they must still be kept within bounds.

Table 2 – Variation of Inmarsat Elevation/Azimuths

Satellite Location	Azimuth		Elevation	
	Max	Min	Max	Min
Atlantic East	200.3	199.2	31.7	27.2
Atlantic West	242.8	240	14.6	10.4
Indian Ocean	108.8	110.9	7.9	5.2

6. Satellite Timing

As already pointed out, in the case of GPS and similar systems using one-way ranging, stability of timing in the satellite is of the utmost importance. These systems commonly carry several atomic frequency standards in each satellite for the derivation of timing. Other satnav systems, using the satellite only as a transponder, maintain all timing at a master control station on the ground and do not require accurate timekeeping in the satellite itself. The only requirement is that the time the satellite transponder takes to turn the signal around and re-transmit it is kept constant.

In one-way systems the monitor station measures the absolute time offset of the transmitted ranging markers relative to whatever is adopted as the system standard time. This in turn means that that time standard itself must be very stable but with the almost unlimited facilities that are available at ground stations this is not difficult.

It is interesting that one of the corrections that must be made is for relativistic effects. Satellites are in a slightly different gravitational field than is the surface of the Earth and are travelling at high speeds. This results in a slight shift in timing and radio frequency that must be compensated and for GPS it means that there has to be an offset of 4.45 parts in 10^{-10} in its internal timing to produce the right signal timing by the time the signal has got to Earth. This is quite a different effect from Doppler-shift.

7. Satellite Management Data

Besides measuring the ephemeris and timing offsets, monitors also keep a watch on general satellite performance. A radio channel separate from navigational transmissions is used to transmit 'housekeeping' data that carries information about the electronic performance of the satellite itself. Several thousand data points are monitored internally in the satellite and most malfunctions can be seen and diagnosed. The satellite itself performs digital monitoring on its own data stream, as does any computer, and can spot bit parity errors and so on, but some malfunctions, such as a slow loss of stability of timing, cannot be monitored on-board and can only be seen by comparison with an external standard.

The criterion of whether a malfunction has occurred is whether the problem affects navigational capability relative to the system specification, and this in turn is a matter of what the user's operational commitment requires. A military requirement linked tightly to a particular operational scenario may differ greatly from that which is needed by a civil user. If the system is a military one then some malfunctions acceptable to the military user may not be to a civil user, or vice-versa. This may require there to be separate monitoring systems for each class of user.

Some malfunctions will cause the satellite to issue its own 'health' warning without needing any ground-station command, while others can be controlled from the ground. These are usually also sent on the standard data stream and can be detected by the user's receiver.

8. Data Transmission to the User

For real-time, that is, navigational purposes, this is universally carried out by using the navigation satellite transmission itself. An upload station carries out the task of uplinking the data stream calculated and formatted by the master control station. It may either be a specialist uplink station or it might be combined with the master control station.

While this is the best way when the system is being used for real-time navigation, eliminating the need for a separate radio signal, it is not the only way, particularly when the system is being used for purposes not requiring real-time results. In commercial systems where the object is to provide vehicle location to a central control room and there is no requirement for on-board position calculation (i.e. non-navigational systems) there is no need to send ephemeris data over the satellite link. All positioning computations are made at the master station where the satellite ephemeris is also calculated. The only necessity is for the satellite link to carry precise timing in some form or other, although it may also carry ordinary communications and vehicle identification information.

9. Integration with Ground-based Systems

Provided all transmitters have their timing locked to the same time-base, a radio-navigation system does not need all its signals to come from the same type of transmitter. Satellite transmitters could be combined with ground transmitters, and there are proposals to do this. One is to put small GPS-like transmitters on the ground in areas where GPS accuracy needs to be reinforced, e.g. for aircraft airfield approach purposes. They would transmit signals indistinguishable from satellites (apart from coding) and therefore receivable on normal GPS receivers without modification. They are called 'pseudo-lites' and are discussed in Chapter 8. Another is to use the existing ground transmitters for the Loran-C, Decca, or other systems in combination with satellite systems. Properly chosen, they could reinforce local accuracy by improving geometry but they would also re-introduce problems of interference, sky wave, and so on. The reinforced system would only be usable out to the range of the ground transmitters, but it has some attractions where there is already a heavy investment in them.

10. Accuracy

10.1 Propagation – Refractive Effects
The frequencies used for satellite navigation vary from the 1225/1575 MHz of GPS/GLONASS to the 11 GHz of Euteltracs. All propagate basically in

straight lines and are not subject to the extensive ionospheric reflection suffered by frequencies below 30 MHz, but there are refractive effects caused both by the ionosphere and the atmosphere.

10.1.1 Ionospheric Refraction

For about 97.5% of their travel, radio signals from a system like GPS are travelling in a vacuum and there are no propagation effects that might alter the time relationship between speed-of-travel and distance. But for the remaining 2.5%, they traverse ionised layers that introduce a slight delay and therefore a small range error. This timing error is due to an actual delay in transmission through the ionosphere, not to 'bending' as is sometimes thought. (Although there is indeed slight 'bending', the geometric change in path-length is so small as to be negligible). The delay depends on the actual radio frequency in use and rapidly gets smaller as frequency increases, roughly in proportion to $\sqrt{f}$. Because of this frequency-dependence, ionospheric refraction can be measured directly if two radio frequencies are used that are phase synchronised and sufficiently far apart in frequency. For instance, Transit uses 150 and 400 MHz (ratio 3:8) and GPS/GLONASS 1225 and 1575 MHz (ratio 12:15) for this purpose. The very small difference in time of reception of the two frequencies is measured and a correction can be calculated. Alternatively, if the density of ionisation is known, for instance from ionopheric sounding stations, the error can be almost exactly calculated and the GPS message has provision for an iono correction term designed to be used by single-frequency receivers to make a partial correction.

It is not as good as the two-frequency method because the correction term has to be averaged over a large area. Systems like WAAS – see Chapter 8 – have more comprehensive arrangements and issue more detailed corrections over their own data-link. The data collected by iono sounding stations is useful for the post-processing done by surveyors because there is more time for detailed analysis. Surveyors who want the extra accuracy in real-time sometimes use receivers that phase-track the L2 frequency which is sufficient to get an iono correction. In any case, ionospheric refraction errors even if uncorrected are smaller than the deliberately introduced Selective Availability error in GPS.

Most civil systems using geostationary satellite relay (Geostar, etc) cannot perform real-time correction because two phase-coherent frequencies are not usually available from ordinary communications satellites. It usually doesn't matter very much because they are not designed for very high accuracies and anyway at the frequencies they use (4–11 GHz) ionospheric refraction is quite small and hardly worth correcting. In the case of Starfix, where high-accuracy surveying is the aim, the mode of operation is essentially that of a closed-loop differential system which automatically allows for refraction effects provided the ionosphere is reasonably stable and constant.

If ionospheric delay were to be the same for all the satellites being received,

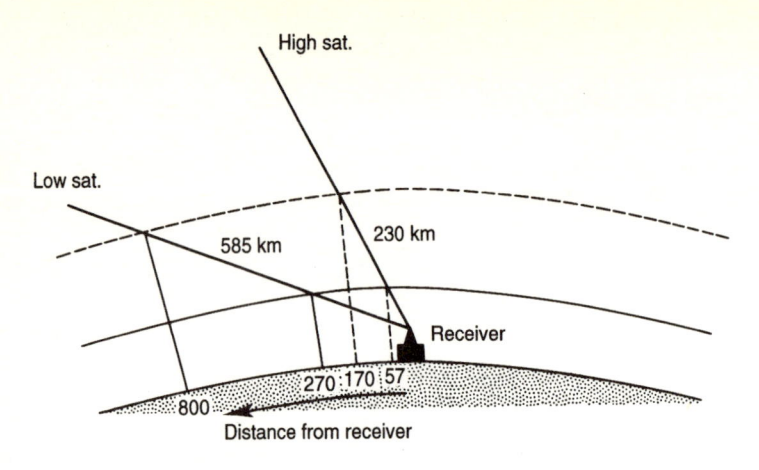

The Air Pilot's Guide to Satellite Positioning Systems

Fig 22 – Signals Traversing Ionosphere

there would be no effect on the fix accuracy. To the receiver it would look exactly the same as an additional clock offset and being common to all satellites would be taken out of the fix computation in the same way. In practice this is not often exactly the case although a large proportion is in fact taken out in this way. The reason why the delay is not the same on all satellites is that the amount of delay depends on how long the signal spends in the ionised layer and this in turn depends on its angle of arrival. For instance, (Fig 22) a downcoming signal at 20° traverses the upper ionospheric levels about 800 kms away from the receiver and the lowest levels about 270 kms away, while a 60° signal will do so at 170 and 57 kms respectively.

The 20° signal is in the ionosphere for about 585 kms while the 60° one is there only for 230 kms, so we might expect the 20° signal to be delayed about 2.5 times more than the high one. It is actually not quite as bad as this because of the curvature of the Earth and other factors and the theoretical figures are given in Fig 23. Note that these are NOT fix errors, only individual ranging errors.

What this diagram shows is that the mean refraction effect for satellite elevations of between 10° and 80° is about 25 metres. Something like this will therefore be taken out of the individual pseudo-ranges for all tracked satellites if they are fairly evenly spread in elevation over this range, leaving errors of perhaps 10–20 metres remaining.

On this basis, for maximum fix accuracy a set of four satellites all at the same elevation should be chosen, but there are other factors. This assumes the ionosphere is absolutely constant in depth and density over the whole area which is hardly ever the case even under sunspot minimum conditions. Fig 24 shows the normal variation in density around the Earth, showing the different densities North/South and East/West.

Clearly any reasonable spread of satellites in azimuth will involve their signals penetrating the ionosphere at widely separated points that may have

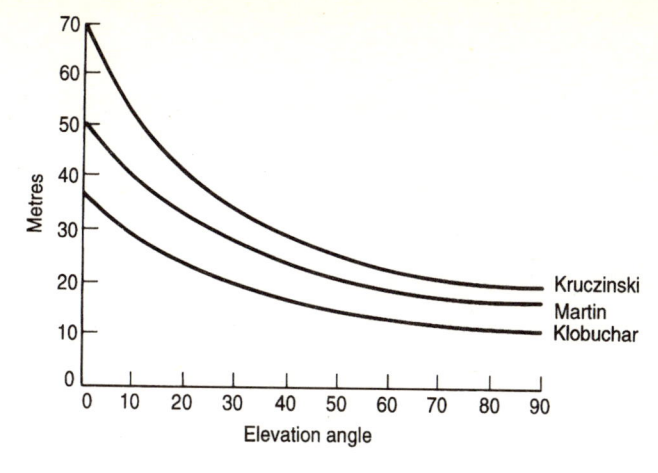

Fig 23 – Theoretical Ionospheric Delays

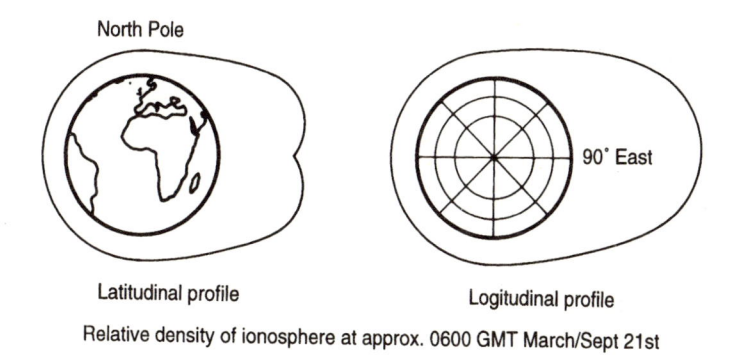

Relative density of ionosphere at approx. 0600 GMT March/Sept 21st

Fig 24 – Variation in Ionospheric Density World-wide

quite different ionospheric densities, particularly at dawn or dusk. Another factor is that there are quite localised and patchy variations in density, known as 'sporadic-E', which are quite prevalent in summer daytime.

Yet another ionospheric effect is that of scintillation which, when it occurs, can cause considerable problems. The major effect is to cause very rapid deep fading, to the extent that GPS signals can totally disappear and of course a receiver can do little about it. Fortunately it occurs only over comparatively small areas; at certain phases of the sunspot cycle; (at sunspot minimum it hardly ever occurs), and for short time periods. Fig 25 shows its geographical and time distribution world-wide.

Really deep fading occurs only when a satellite signal actually traverses a scintillation zone. These zones are fairly localised and because satnav has to use a geographically widely dispersed set of satellites the chances are that only one or two satellites will be affected. The receiver would then simply

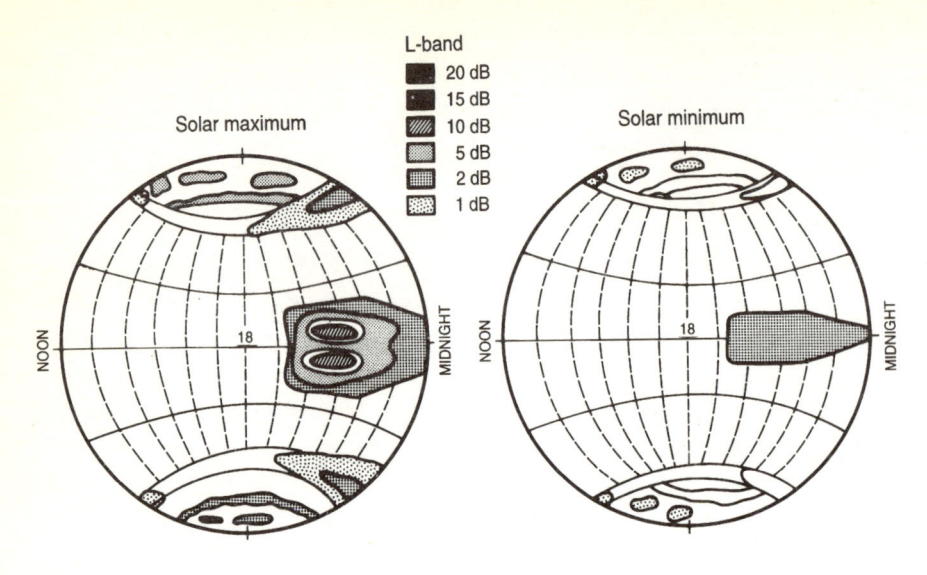

The Air Pilot's Guide to Satellite Positioning Systems

Fig 25 – Scintillation

choose another satellite and the total effect might be simply a temporary lowering of accuracy due more to the less favourable geometry than anything else. Where it might be of more significance is when geostationary satellites are being used either for wide-area augmentation or for primary navigation and their signals have to cross a disturbed zone. In that case they might be lost for significant periods.

10.1.2 Atmospheric Refraction

This is not dependent on radio-frequency and is roughly the same at all frequencies. It is caused mainly by water vapour in the atmosphere (*not* by liquid water such as rain) and can be quite accurately calculated if local meteorological measurements are available. Fig 26 shows its effects and it can be seen that they are considerably smaller than the ionospheric effects.

As with ionospheric refraction, if the same error were present on all the received satellite signals then it would not cause a fix error, and because the troposphere is much lower and not as thick as the ionosphere there is more chance of this being so. The top of the troposphere is for this purpose at about 20 kms or so, so a 20° elevation satellite signal would travel through this level about 55 kms from an observer and therefore it only has to be of constant refractive properties over a comparatively small area.

There are still the differing amounts of refraction due to different satellite elevations to be considered, but it can be seen from Fig 26 that like iono delays about half the maximum error is present at all elevations. This is about ten metres and since it will be common to all satellites it will be removed and might leave about ten metres uncorrected at 10° elevation. On average, this

Basic Satnav Systems

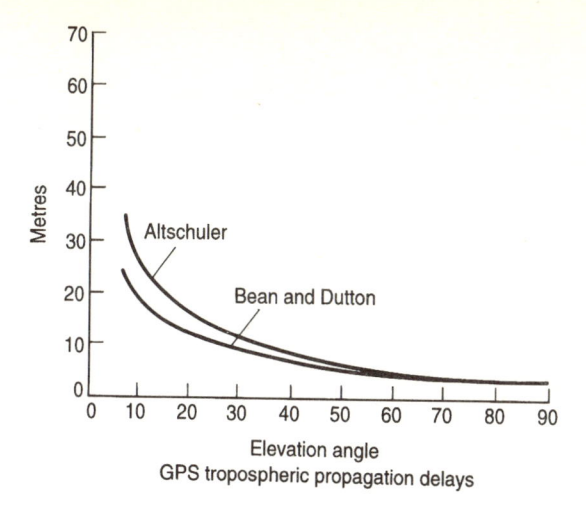

Fig 26 – Atmospheric Ranging Errors

error will only be present on one satellite, so the resultant fix errors are usually quite small – a few metres – and navigation receivers do not usually attempt any correction. High-accuracy survey receivers may do so by using inputs of local pressure, temperature and humidity, enabling an almost complete correction.

Most tropospheric refraction occurs in the last few thousand feet closest to the ground where air density (and, often, water vapour concentration) is greatest, so fix errors are even less for aviators flying at anything other than low altitudes. Aircraft at FL30 and above can safely ignore this error.

10.2 Propagation – Multipath

The other propagation limitation on accuracy is multipath, where the satellite signal is reflected from one or more surfaces and the receiver gets several signals arriving at slightly different times (Fig 27). The delay between them confuses the receiver timing circuits and produces rather noisy timing.

Reflections occur from land or sea surfaces, the wings or other parts of an aircraft, or nearby objects when on the ground. They arrive from different directions but because the receiver aerial has no directivity it cannot discriminate between them. Not much can be done about this other than careful aerial siting but its effects should not be over-estimated. The conditions required for a surface to reflect a GPS signal in a particular direction are quite critical and with the continuous movement of both a satellite and an aircraft, multipath is often fleeting and hardly noticeable. It is more serious for land surveyors operating from a fixed site. Development work is proceeding on receiver circuits that can recognise when multipath is present and either switch to another satellite temporarily or briefly suspend timing.

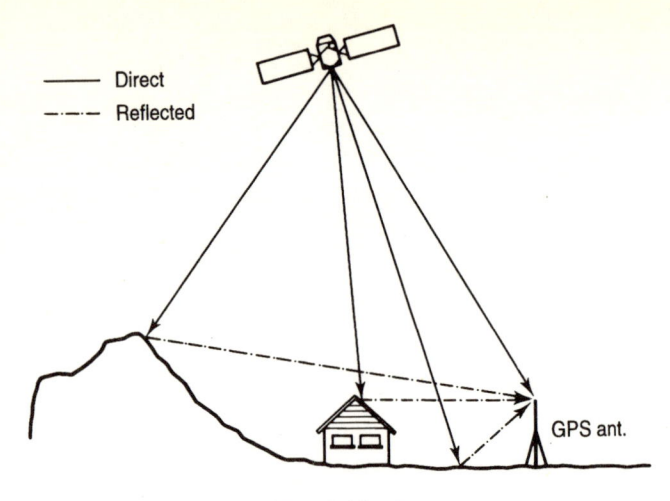

The Air Pilot's Guide to Satellite Positioning Systems

Fig 27 – Multipath

10.3 Geodetic Transformation Errors

Range from an Earth-bound radio transmitter can be drawn on a chart as an arc of a circle but it can only be drawn like this if it is assumed that both the user and transmitter are on or very near to the surface of the Earth, which is usually the case. Thinking three-dimensionally, the Earth's surface provides the second sphere necessary to turn the first one into an arc of position. Errors in the radius of this second sphere do not matter much provided it is the same one that was used for drawing the map, which of course it will be. This might seem a rather laboured way of describing how to draw a position line but it becomes very relevant when the transmitter is on a satellite. Satellites are certainly not on the Earth's surface and their 'height' can never be ignored. Although satellite 'height' (semi-major axis, SMA) can be calculated quite accurately from observations of orbital period, it is relative to the mass whose gravitational attraction makes them stay in orbit; in other words, the mass-centre of the Earth. The result is that the altitude of the satellite above the surface of the Earth cannot be derived directly from Keplerian parameters – it must be calculated by subtracting from its SMA the distance of the Earth's surface from its mass-centre, which obviously cannot itself be directly measured. Usually, to make things simple, an average figure for the Earth's radius (6,378.8 kms) is used, but this is hopelessly inaccurate for precise navigation. The converse process, subtracting satellite altitude (which can be derived from range) from SMA is much easier and gives the distance from the local surface of the ranging station to the mass-centre ('d') (Fig 28). To put this on a map, the mass-centre and 'd' must then be reconciled with the centre and radius of whatever spheroid was used as the basis for the map – and this is where the trouble starts!

Basic Satnav Systems

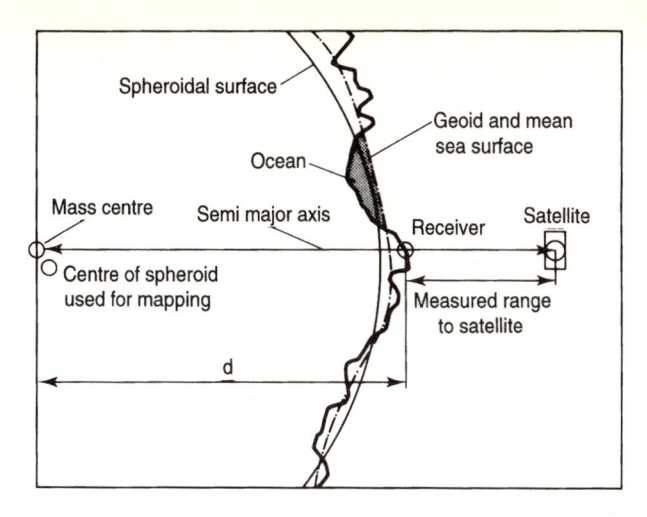

Fig 28 – Height and Altitude

If 'd' could be assumed to be the same world-wide, a figure of the Earth would be obtained that would be a perfect spheroid of radius 'd'. It would have little relationship to the real world except near the ranging station because the Earth is far from being a perfect sphere, but it would be mathematically simple for local mapping calculations. Other ranging stations would obtain different values for 'd' resulting in different spheroids; obviously highly unsatisfactory for a world-wide system.

10.3.1 The Geodetic Problem

Early mapping surveyors measured distances and directions between objects they could see and drew their maps accordingly, assuming a flat Earth. Later, as bigger areas were covered, it became harder for them to make their measurements tally because of the Earth's sphericity which could no longer be ignored. In order to take it into account, they developed mathematical models of the Earth's shape but because they still did not know accurately its true overall shape and it was difficult to handle a non-spherical shape, they assumed it was spherical and calculated the best shape (spheroid) to fit their observations. This led to different surveyors in different parts of the World adopting different spheroids (Fig 29).

Local maps were then drawn on the basis of whatever local spheroid had been calculated. Politics usually ensured that each country adopted its own and didn't worry too much about what its neighbours used. When radio navaids came along their positions were given in terms of the local map and as long as they had rather short ranges it didn't matter very much except where the maps of one country abutted those of another.

Satellites don't confine themselves to a single country, so the spheroid used

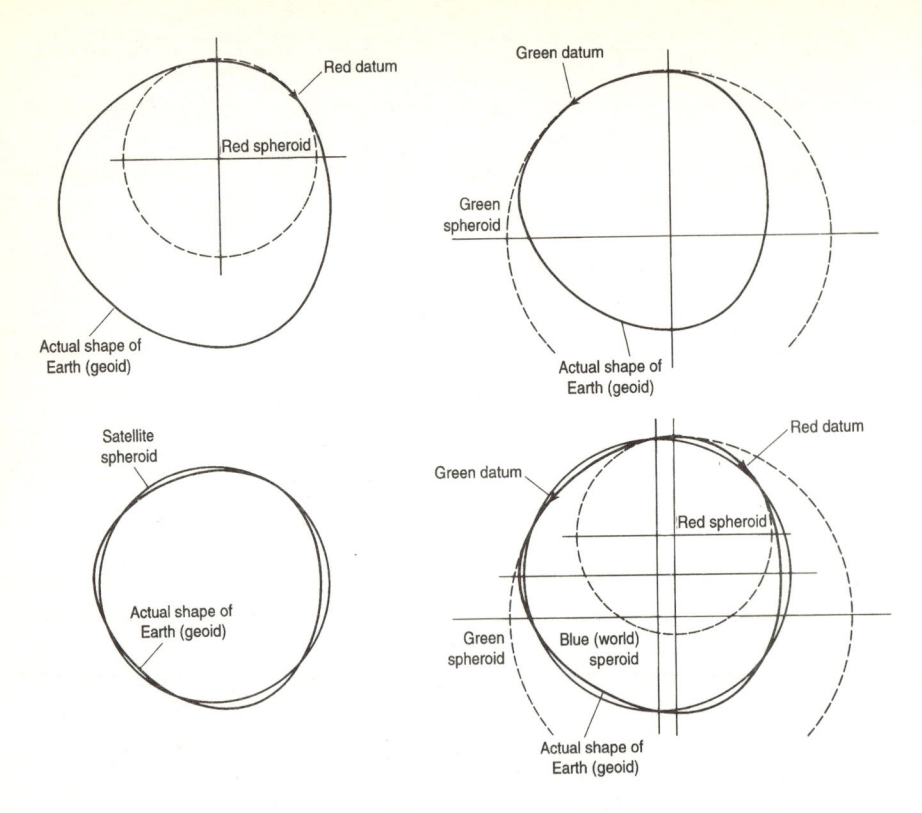

Fig 29 – Differing Spheroids

to convert satellite fix measurements back to the Earth's surface has to be a world-wide average – in fact, the average true figure of the Earth. Thus, positions obtained from a satnav system are basically produced in a co-ordinate system based on a spheroid that does not match anything used for existing maps and charts. The first satnav system, Transit, was used extensively by surveyors for re-determination of the Earth's figure and as a result the differences between most datums and spheroids have now been established. It doesn't help the navigator much because nearly all charts and maps are still being produced using old spheroids. The ideal would be to draw new maps and charts using the satellite spheroid, but one estimate is that the conversion of all the current maps and charts will not be completed until the middle of the next century.

One way of managing the problem is to put a note on maps and charts to the effect that latitudes and longitudes obtained in the satellite datum and spheroid should be moved by so many metres to correspond with the chart. It is only an average figure over the whole chart and some inaccuracies will remain.

Basic Satnav Systems

An easier way is to do it in the satnav computer. Many now have the facility of outputting satnav position ready converted to any of the common spheroids and datums. A user doesn't have to know anything about spheroids or datums – usually it is just a matter of entering the country whose charts are in use. But, like the previous method, the constants used are often just an average and there are still some residual small errors. Note that although this converts the satellite positions into the local datum, it cannot correct for any mapping errors due to surface features having been incorrectly surveyed in the first place.

Some commonly-used datums and spheroids and the differences between them and the satellite datum and spheroid are listed in Appendix 4.

A description of what has become known as 'the WGS–84 problem' is given at Appendix 5, with special reference to the situation in the UK. Also in this Appendix is the UK Ordnance Survey statement of policy regarding re-mapping.

Chapter 3

The Global Positioning System – GPS/Navstar

1. The Defense Navigation Satellite System, Timation and Project 621B.

In 1961 studies were put in hand in the USA for a satellite navigation system that would be suitable for both marine and air use, under the title of 'The Defense Navigation Satellite System'. Although the 'Transit' system was already in partial operation, its limited usability, particularly for aircraft, was recognised and the need for a more comprehensive system was seen. There were two main objectives for the new system:

1. Precision Weapon Delivery
The number of weapons/aircraft required to ensure a target is hit varies approximately as the square of the delivery accuracy, so that doubling accuracy reduces the effort required to a quarter. The knock-on effects in aircraft procurement, training and support are considerable and the savings that were thought possible more than justified the cost of a satellite system.

2. A Reduction in the Number of Navigation Systems
Because of the inability of any one existing system to provide high accuracy, long range, and instant availability anywhere operations might be required, there had been an explosion in the number and types of military positioning systems, all requiring specialist support and training. A single system capable of fulfilling all objectives would enable considerable savings.

As its contribution the US Navy undertook in 1964 a project known as Timation to demonstrate the use of very precise synchronised clocks in space, an essential prerequisite for a navigation system. Two satellites for this system, Timations I and II, were flown in 1967 (1967-53F) and 1969 (1969-82B). They were succeeded by Navigation Technology Satellite One (NTS-1) in 1974 (1974-54A) carrying a rubidium atomic clock, and NTS-2 in 1977 (1977-53A) carrying a caesium clock (see Table 3). Although no longer operational they are still in orbit and occasionally transmit. NTS-2 was later designated the first NAVSTAR GPS Phase 1 satellite. There was to have been an NTS-3 carrying a hydrogen maser clock, but it was never flown.

The Global Positioning System – GPS/Navstar

Table 3

Satellite	T-1	T-2	NTS-1	NTS-2
Launch Date	31.5.67	30.9.69	14.7.74	23.6.77
Altitude (km)	920	920	13,620	20,200
Inclination	70	70	125	636
Eccentricity	.001	.002	.007	.0004
Weight (kg)	39	57	295	440
Power (W)	6	18	125	400
Frequencies	UHF	VHF/UHF	UHF/L	UHF/L
Oscillator	Quartz	Quartz	Qtz/Rb	Qtz/Cs
Drift (10^{-13}/day)	300	100	7	1.5

In parallel with the Navy, the US Air Force conducted a series of system design studies known as Project 621B, for a highly accurate three-dimensional navigation system, which included experiments with transmitters located on mountain-tops to simulate satellites. The main purpose was to validate what was the then new concept of using pseudo-random noise modulation (PRN) for ranging purposes. Initially, several constellations of satellites in highly eccentric 24-hour orbits were envisaged for the operational system, but it became clear that the large Doppler-shifts inherent in this type of orbit would lead to receiver design complications and the idea was abandoned. This initial design also had all satellites being continuously uploaded with timing and orbital information from ground stations but this was militarily insecure and was also abandoned. Instead, the on-board highly accurate atomic clocks being used for Timation were adopted as the basic timing method for the new system.

These two activities were integrated by a memorandum of the Deputy Secretary of Defense of 17 April 1973, and final approval to proceed with the new system was obtained on 17 December 1973. The Air Force was designated as the Executive Service and the system re-designated Navstar, the Global Positioning System. How this title came into being is interesting. The 'GPS' name was the suggestion of Gen. H. Stehling, the then Director of Space for the US Air Force, who pointed out that using the word 'navigation' as in the original title was incorrect. Then, it was suggested by a director in the Pentagon who had control of funding that 'Navstar' would sound nicer, so in order to retain support from both departments it was renamed 'Navstar – the Global Positioning System'. Col. Brad Parkinson, who was in charge of the GPS office at the time, has recorded that 'Navstar' was not an acronym for anything and only a nice-sounding name. However, later it was discovered that the TRW company had proposed a navigation system that they had termed 'NAVSTAR' as an acronym for 'NAVigation System Timing And Ranging', which caused some confusion. The title Navstar now appears to be falling into disuse, not having appeared on any of the recent US civil documents dealing with GPS, although it is still seen on military documents.

2. Launch History

GPS satellites, like all other satellites, are allocated an international scientific designator of the form 'year; number in year of the launch; piece of launch' e.g. 1987–26b, indicating the second object that was launched by launch number 26 in 1987. This designator is not widely used when referring to GPS satellites except by the scientific community, the US DoD listing them serially in order of production as 'Navstar' numbers, and then by reference to the PR code they use (see Appendix 7 for a description of what 'PR code' means). There are thirty-two PR codes in use for GPS but many more are possible and some of these others will be used in the Inmarsat augmentation system (see Chapter 7). Receiver manufacturers and users have found it more useful to refer to satellites by their PR number (PRN) so it has become general practice in the navigation world to use these numbers rather than anything else. The Block I satellites had different 'Navstar' and 'PR' numbers, which led to some confusion, so in the Block II units this was rationalised as far as possible by allocating the same numbers for both 'Navstar' and 'PR'. This could only be maintained up to PRN 32, since the system only allows for 32 PRN numbers, so subsequent satellites have had to revert to different Navstar and PR numbers.

2.1 Block I (Test and Development) Satellites

As already mentioned, the first GPS navigation signals were broadcast from NTS-2 in June 1977, continuing until it failed in February 1978. The construction of four pre-operational GPS satellites, known as the Block I series and built by Rockwell, had been authorised in 1974, the first launch being Navstar 1 (1978–20A), followed rapidly by Navstars 2 (1978–47A), 3 (1978–93A), and 4 (1978–112A). Two more satellites were then authorised and in 1980 5 (1980–11A) and 6 (1980–32A) were successfully launched. By then, six more had been ordered, but Navstar 7 was destroyed on lift-off through a launcher failure in Dec. 1981. The others were launched between 1983 and 1985. One of the early problems was that the lamps in the on-board rubidium oscillators failed quickly, a problem overcome from Navstar 5 onwards. From Navstar 4 on, one or more caesium clocks were flown in each satellite, proving very satisfactory after initial power unit problems were solved. At one time, there was a plan to fly hydrogen maser clocks, (as early as NTS-3), but development problems caused this to be dropped, and there is no current plan to resurrect it. It was thought at the time that the more stable the on-board clock was, the better, but the development of computer software modelling techniques has now reduced the need for ultra-stable clocks. These satellites were destined purely for demonstration and development purposes and designed for only a limited lifespan (5 years), but all except one have handsomely exceeded this figure.

Initial testing was done using the ground transmitters available from the

The Global Positioning System – GPS/Navstar

Fig 30 – GPS Satellite

Project 621B trials and linking real satellites with them as they were launched. When four satellites were in orbit the ground system was decommissioned but the concept of augmenting satellites with ground transmitters lives on in the pseudo-satellite ('pseudo-lite') idea for landing aircraft.

Table 4
Block I – (Test and Development) Satellites

SVN	PRN			
1	4	Launched 22 FEB 78;	removed from service	17 JUL 85; life 7 yrs.
2	7	Launched 13 MAY 78;	removed from service	12 FEB 88; life 10 yrs.
3	6	Launched 06 OCT 78;	removed from service	18 MAY 92; life 14 yrs.
4	8	Launched 10 DEC 78;	removed from service	14 OCT 89; life 11 yrs.
5	5	Launched 09 FEB 80;	removed from service	11 MAY 84; life 4 yrs.
6	9	Launched 26 APR 80;	removed from service	06 MAR 91; life 11 yrs.
7		Unsuccessful launch 18 DEC 81		
8	11	Launched 14 JUL 83;	removed from service	04 MAY 93; life 10 yrs.
9	13	Launched 13 JUN 84;	removed from service	28 FEB 94; life 10 yrs.
10	12	Launched 08 SEP 84;	still operating, 1995.	(Life so far 10.5 yrs)
11	3	Launched 09 OCT 85;	removed from service	16 FEB 94; life 9 yrs.

Note: 'Removed from service' usually occurs because the frequency standards have deteriorated so badly that timing is insufficient for navigational use. The other satellite systems may still be working quite well.

The Air Pilot's Guide to Satellite Positioning Systems

2.2 Block II (First Operational Series)

The Block II satellites were designed as the first operational series and are considerably bigger and heavier than the Block Is – 846 kgs v. 450 kgs. They incorporate the ability to selectively provide different levels of accuracy as well as many detail improvements, including a design life of seven years as against the five years of Block I. In the interests of lower costs a single contract for the building of twenty-eight satellites was awarded to Rockwell in 1984. The original intention was that launches would be made using the Delta launch vehicle but in 1979 this was changed to the Shuttle vehicle and the BIIs were designed with this vehicle in mind. Although satellite manufacture kept to schedule, the 1986 *Challenger* Shuttle accident meant that the launch schedule slipped badly and it was decided in 1987 to launch Block IIs only on the revised McDonnell Douglas 'Delta II' launcher. Table 5 gives the Block II launches up to early 1995 – so far all have been successful.

Table 5
Block II – Individual Satellite Status

SVN	PRN			
13	2	Launched 10 JUN 89;	usable 10 AUG 89;	operating on Cs std
14	14	Launched 14 FEB 89;	usable 15 APR 89;	operating on Cs std
15	15	Launched 01 OCT 90;	usable 15 OCT 90;	operating on Cs std
16	16	Launched 18 AUG 89;	usable 14 OCT 89;	operating on Cs std
17	17	Launched 11 DEC 89;	usable 06 JAN 90;	operating on Cs std
18	18	Launched 24 JAN 90;	usable 14 FEB 90;	operating on Cs std
19	19	Launched 21 OCT 89;	usable 23 NOV 89;	operating on Cs std
20	20	Launched 26 MAR 90;	usable 18 APR 90;	operating on Cs std
21	21	Launched 02 AUG 90;	usable 22 AUG 90;	operating on Cs std
22	22	Launched 03 FEB 93;	usable 04 APR 93;	operating on Cs std
23	23	Launched 26 NOV 90;	usable 10 DEC 90;	operating on Cs std
24	24	Launched 04 JUL 91;	usable 30 AUG 91;	operating on Cs std
25	25	Launched 23 FEB 92;	usable 24 MAR 92;	operating on Rb std
26	26	Launched 07 JUL 92;	usable 23 JUL 92;	operating on Cs std
27	27	Launched 09 SEP 92;	usable 30 SEP 92;	operating on Cs std
28	28	Launched 10 APR 92;	usable 25 APR 92;	operating on Cs std
29	29	Launched 18 DEC 92;	usable 05 JAN 93;	operating on Cs std
31	31	Launched 30 MAR 93;	usable 13 APR 93;	operating on Cs std
32	1	Launched 22 NOV 92;	usable 11 DEC 92;	operating on Cs std
34	4	Launched 26 OCT 93;	usable 22 NOV 93;	operating on Cs std
35	5	Launched 30 AUG 93;	usable 28 SEP 93;	operating on Cs std
37	7	Launched 13 MAY 93;	usable 12 JUN 93;	operating on Cs std
39	9	Launched 26 JUN 93;	usable 20 JUL 93;	operating on Cs std
40	6	Launched 10 MAR 94;	usable 28 MAR 94;	operating on Cs std

(Launch dates have not been transposed out of sequence – some satellite/launcher combinations were delayed after having been designated).

The Global Positioning System – GPS/Navstar

By 1995 the entire GPS constellation will be made up entirely of Block II satellites. If any of the Block I units are still serviceable they will be switched off and taken out of use, not having full military operational capabilities.

Some of the later Block II satellites, designated Block IIA, incorporate changes in the navigational payload and have some additional facilities.

2.3 Block IIR series.

Block IIR (Figs. 31/32) are the replacements for Block II, with several new features. A contract for twenty, with options for six more, was let to Martin Marietta (formerly GE-Astro) in June 1989, for deliveries starting in 1995. Amongst their new features will be the ability to function for up to 180 days without upload from the ground, and increased radiation hardening. In order to achieve the 180-day autonomous capability, they will be able to communicate with each other for orbit-determination purposes.

Current intentions (late 1994) are that several may be launched in the 1996-1997 time frame for test and qualification purposes, but no large-scale launch programme will be initiated until the failure of BII units makes it essential.

The procurement of these BIIR satellites means that the future operation of GPS is assured and even if satellites fail exactly at the end of their design life

Fig 31 – Block IIR

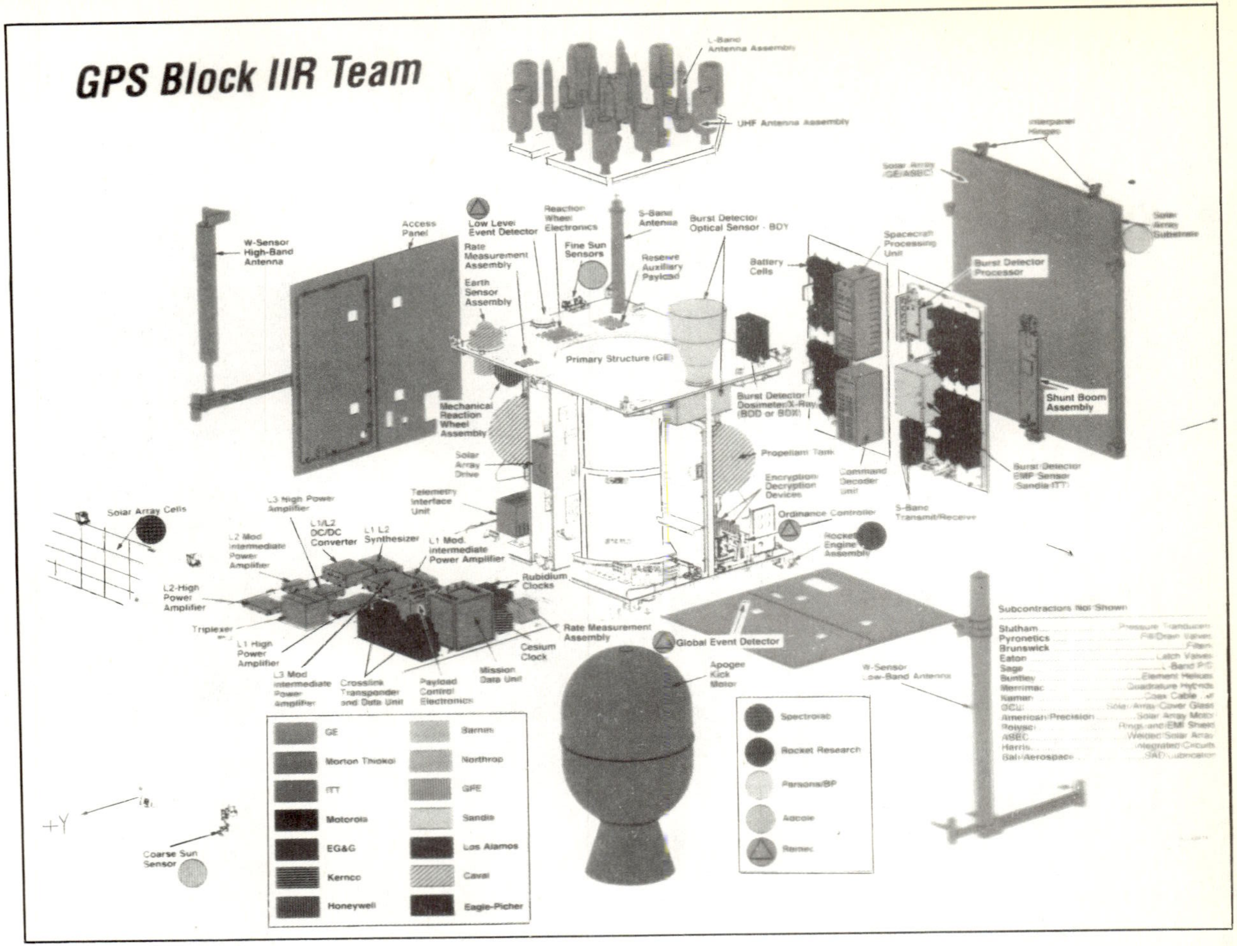

Fig 32 – B**II**R components

of seven years the system will remain operational until at least 2006. What happens after that has not yet been determined, although going by past history many of the satellites will exceed seven years life.

3. System Operation

3.1 Orbital Patterns

For optimum accuracy, a series of satellites spaced around several different planes are required, as described in Chapter 2. US Department of Defense studies showed that the most economical way of achieving their mission requirements was to have either three or six planes with eight or four satellites in each, a total of twenty-four satellites, orbiting in approximately twelve-hour orbits (semi-major axis 26,560 kms).

The Block I satellites were put into nearly circular orbits at 63° inclination, close to the optimum figure of 63.4° for minimising the drift of perigee, (important for maintaining good geometry with only a few satellites) but the Block IIs have used 55°, originally to suit the Shuttle launch but now said to be more economical in launching and station-keeping costs.

The exact pattern of planes and orbital positions within them has changed a number of times, but the overall pattern is now for six planes at 60° longitudinal intervals, with satellites spaced round each plane at either 110° or 40° intervals. The pattern is irregular because the major consideration is to maximise accuracy in certain areas and a regular pattern is not the best way to do it. The current orbital pattern is shown in Fig. 33; Fig 34 shows the actual positions relative to the Earth true to scale. Each plane has multiple lines because the actual orbits are shown, which differ slightly from each other.

3.2 Transmitted Signals

Two frequencies, 1227.6 (L2) and 1575.42 MHz (L1) are used in parallel for the primary navigational signals and are radiated continuously. Other frequencies of 1381.05 and 2227.5 MHz are also transmitted intermittently but not used for navigation, although it is interesting to note that 1381.05 has an exact phase relationship to the two navigation frequencies – they are all derived from 10.23 MHz (×120, ×135 and ×154). Phase coherence is essential for navigation signals.

The modulation method used is pseudo-random noise spread-spectrum (PRNSS), providing accurate timing and other advantages in the rejection of interference both deliberate and natural. PRNSS permits all satellites to use the same radio-frequencies, individual satellites being identified by their differing PR codes. In orbit, they will also have different Doppler-shifts. A full description of how the GPS radio signal is derived is given in Appendix 7, but a shorter description follows.

The GPS spread-spectrum signal is achieved by using high-speed digital

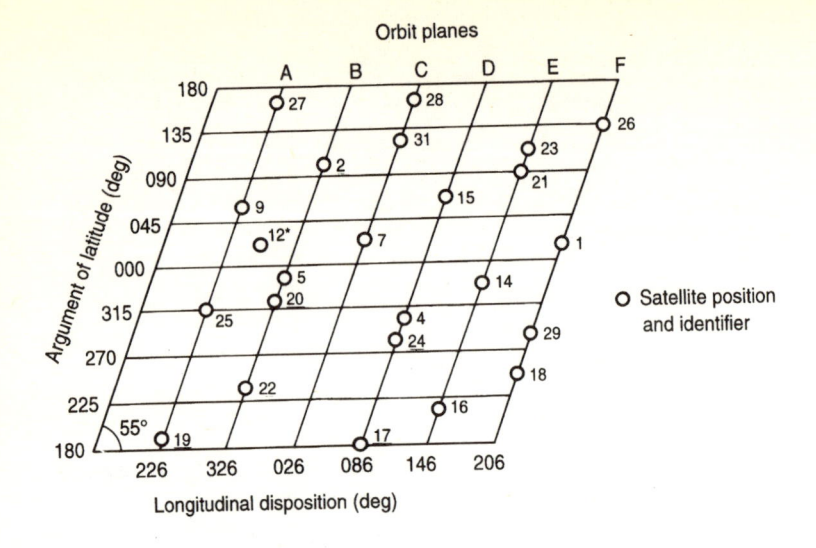

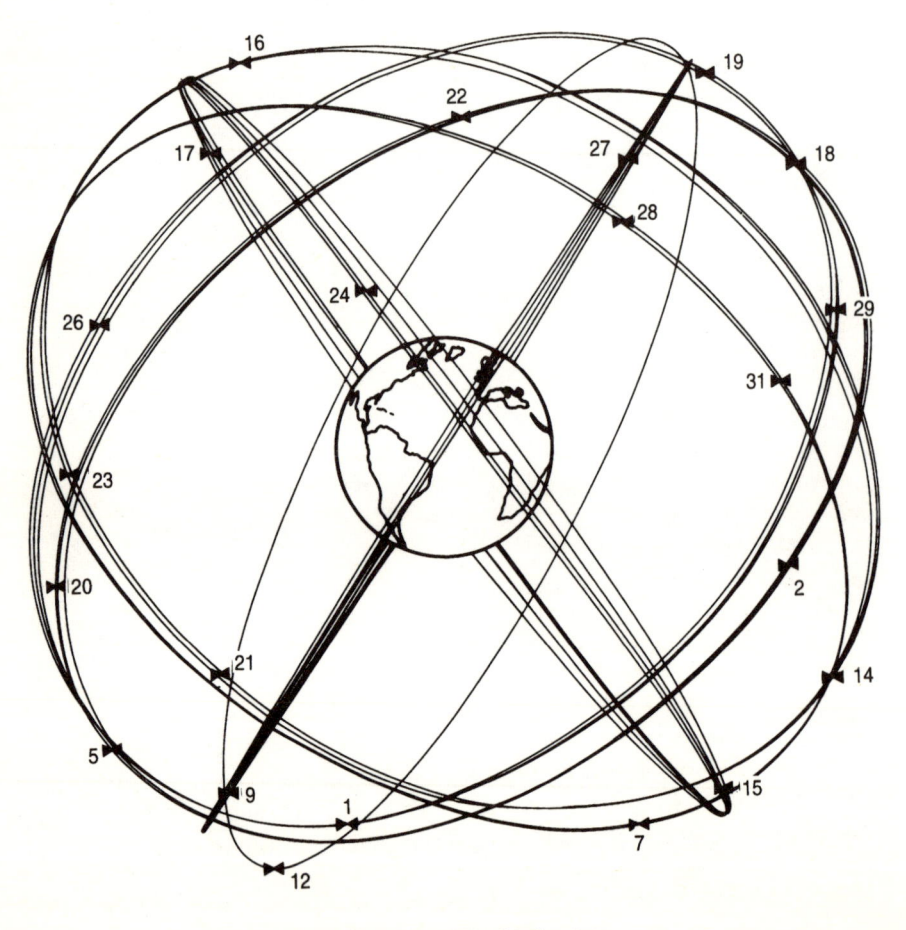

Figs 33/34 – GPS orbital pattern

sequences to modulate a radio-frequency carrier signal. One signal (known as C/A, coarse/acquisition) is generated using a digital stream switching at a rate of 1,023,000 times a second, resulting in a signal bandwidth of about 2 MHz, while the other (P, or precision) uses a digital stream switching exactly ten times faster, or 10,230,000 times/second, giving a 20 MHz bandwidth.

The L1 radio signal is modulated by both the C/A and P digital streams, while the L2 signal uses only the P code. As a result, civil receivers usually have provision only for the L1 frequency.

Accurate timing is ensured by this process but orbital and other data also has to be transmitted on the same signal. There is no great amount of this data and it can be transmitted by using a slow data rate of 50 bits/second which is modulated on top of the basic high-speed bits by periodically reversing their phase. The high-speed sequences are neither truly random nor regular, hence their name of 'pseudo-random'. They are generated by using a digital code sequence peculiar to each satellite, referred to by the 'PRN' number. The C/A digital sequence is generated using a digital code that repeats itself every one-thousandth of a second making it easy and fast for a receiver to acquire. No attempt is made to further encode this signal.

The P-code is quite different. In the first place, it is derived from a digital code that does not repeat itself for 200 days. This alone would make it very difficult to decode from scratch but in addition the whole code is never sent. Instead, sections that re-start every seven days are used. Lack of knowledge of the point this code has reached would make de-coding either very slow or impossible, even for a receiver that knows the code being used. To get over this, the C/A code signal contains a message (the 'hand-over word', HOW) telling the receiver the stage the P-code has reached.

Then, this information is itself encrypted in order to prevent non-military users using it. Encryption is a form of 'scrambling' the P-code in a way that can only be unlocked by use of a 'key' module in the receiver. When encrypted the P-code is referred to as the 'Y' code.

To recover these signals a receiver must first generate an exactly similar digital stream to the one being used by the satellite. In the case of the C/A coded L1 signal it can do this quite simply by using the PRN number of the satellite it is trying to find. It then uses this synthesized digital stream for demodulation, reversing the wide-band spreading and concentrating the radio energy back into a narrow bandwidth. Provided a sufficiently short clock sequence is used, known to both receiver and transmitter, and they are stationary, this is not difficult. But in the case of GPS, neither may be stationary and there may be quite significant Doppler offset. Also, the time the signal takes to get from the transmitter to the receiver will offset the digital stream even more – this is of course the essential navigational measurement.

All this means the receiver cannot know where in absolute time the incoming signal might be found, and it has to perform a search process by jumping its local code in small steps until coincidence is found. Details of this process are given in Appendix 7.

Because P-code is some ten times faster than C/A code, it was originally thought that its accuracy would also be considerably greater and early estimates were that P-code would provide fixes around the 50-metre region and C/A code around 500 metres. Once the Block I satellites were up it was found that both estimates were wrong, P-code giving about 15 metres and C/A code about 50 metres. This created a problem because while a C/A code accuracy of 500 metres was considered to have no particular military significance, 50 metres certainly did, and as a result the Selective Availability technique was introduced which deliberately degrades the available accuracy (see later).

The specified power levels at the user's antenna, which is assumed to have a nearly hemispherical coverage pattern, are -160 dBW for L1 (1575) and -163 dBW for L2 (1228). Since propagation loss over 26,000 kms is 185 and 182.5 dB respectively, this indicates a satellite power of between 25 and 19.5 dBW. These signal levels provide for a margin above average noise levels of only 6 dB or so, perfectly acceptable under normal conditions but if there is extraneous noise induced by, for instance, incorrectly operating radar transmitters on a nearby frequency then GPS signals can easily be lost.

To a listener using a narrow-band receiver (and enough antenna gain), both signals sound like random white noise and are virtually indistinguishable from the normal receiver background noise, except for the C/A signal carrying a slight but noticeable 1 kHz whine due to the code repetition rate.

3.3 Transmitted Data

The receiver has to be able to calculate precisely where each satellite is and allow for the offset each satellite clock has (the difference between it and system time). These data are received from the satellite via the 50 bps data stream and are uploaded by a ground control station.

As described in Chapter 2, there are several mathematical ways of describing satellite position. For GPS, a pseudo-Keplerian representation was adopted (Table 6). It is called 'pseudo' because it is only an accurate fit to the actual orbit for that part in which it is transmitted and does not necessarily represent correctly the complete orbit. This is done because the orbit of a satellite as described by true Keplerian parameters is a mean over the whole orbit and cannot take account of any minute-by-minute deviations. These small deviations can be sufficiently large to cause errors in the fix computation if not corrected.

In addition to this navigational ephemeris, a true Keplerian ephemeris valid for a considerable number of orbits is transmitted as an 'almanac' to permit approximate prediction of satellite position for receiver acquisition. This almanac is quite accurate enough for all purposes except position determination and can be used for best-DOP satellite selection.

The Global Positioning System – GPS/Navstar

Table 6
Transmitted GPS Satellite Pseudo-Keplerian Elements
(*same names for both navigational ephemeris and almanac*)

Parameter	Units	Remarks
Mean Anomaly	Semicircles	
Mean Motion	Semicircles/sec	(1)
Eccentricity		
Semi-major Axis	Metres	(2)
Right Ascension	Semicircles	(3)
Inclination	Semicircles	
Argument of Perigee	Semicircles	

Remarks:
(1) Not absolute mean motion but a correction to the mean motion that can be calculated by classical methods using Earth's gravitational constant and the semi-major axis of each satellite.
(2) Actually square root of semi-major axis.
(3) Referred to Greenwich meridian at beginning of GPS week and therefore not classical Right Ascension which is referred to the vernal equinox (first point of Aries). This removes the need to calculate GHA Aries. Sometimes referred to as 'longitude of the ascending node' (LAN).

In the navigational ephemeris, there are seven more non-Keplerian parameters transmitted for use as corrections, permitting second-by-second high-precision orbit calculation. A full list is given in Table 7. Also transmitted are data concerning the offset of each satellites' time relative to GPS time. GPS time is basically UTC as it was in 1984 ignoring the leap seconds that have occurred since – it is currently (1994) some ten seconds different from 1994 UTC1 and the difference will get bigger by a second a year if UTC is adjusted by that amount each year.

4. The Operational Control System

The ground control system for GPS consists of a number of world-wide monitors in constant communication with a GPS Master Control Station (MCS) in Colorado Springs, USA. They are located at Ascension Island (Atlantic), Diego Garcia (Indian Ocean), Guam (Pacific), Hawaii (Pacific), and at the MCS itself. A sixth station at Cape Canaveral will be added by the end of 1994. They maintain contact with the MCS via the US DoD TDRSS satellite network and other links (Fig 35). The entire system is known as the Operational Control System (OCS).

Widespread as it is, the OCS can only monitor GPS satellites about ninety-two per cent of the time. This is sufficient for its military role, but not for many

Table 7 – Ephemeris Data Definitions.

M_o	Mean Anomaly.
δn	Mean Motion difference from computed value.
e	Eccentricity.
$(A)^{1/2}$	Square Root of Semi-Major Axis.
Ω_o	Longitude of Ascending Node.
i_o	Inclination.
w	Argument of Perigee
Ω dot	Rate of Right Ascension
i dot	Rate of Inclination Angle
C_{uc}	Arg. of Latitude – Cosine correction term.
C_{us}	" – Sine correction term.
C_{rc}	Orbit Radius – Cosine correction term.
C_{rs}	" – Sine correction term.
C_{ic}	Inclination – Cosine correction term.
C_{is}	" – Sine correction term.
t_{oe}	Reference time.
IODE	Issue of Data (Ephemeris)

civil applications, for which it must be augmented. The MCS is the central node for GPS satellite telemetry and for monitoring the performance of each satellite. If the radio signal or the navigation message from a satellite becomes degraded then it is the responsibility of the MCS to do something about it.

The monitors are automatic and unmanned and track the navigation messages from every satellite as well as measuring range and range rates. They also download housekeeping telemetry via a separate S-band channel and send everything to the MCS for analysis. If a telemetry point is out-of-tolerance it is up to the MCS to decide whether or not the problem affects the satellite's military capability as part of the overall GPS system. Sometimes a fault may be deemed acceptable for military purposes and the satellite left set healthy, although it may be unacceptable for civil use. Also, by the time a fault is detected, reported, analysed and a decision made as to its acceptability, several hours may have passed, although 30–45 minutes is more typical. Civil users may want to have notification much faster than that.

The MCS operates in a basic fifteen-minute cycle. Tolerance and validation checks of the measured, observed, and predicted pseudo-ranges of each satellite are made at this interval. If within specifications they are used to update a system Kalman filter, but if outside the problem is determined and remedied. This may involve an unscheduled data uplink to the satellite involved.

Every day, the MCS sends basic tracking data to the Defense Mapping Agency which uses it in conjunction with its own tracking network to determine a Precise Ephemeris. This is more accurate than the Broadcast Ephemeris, being the result of actual tracking rather than a forecast, but is not available until some days afterwards. It is used mainly by land surveyors and

The Global Positioning System – GPS/Navstar

- ■ Master control station
- ● Monitor station
- ▲ Ground antenna
- ◆ Backup control station

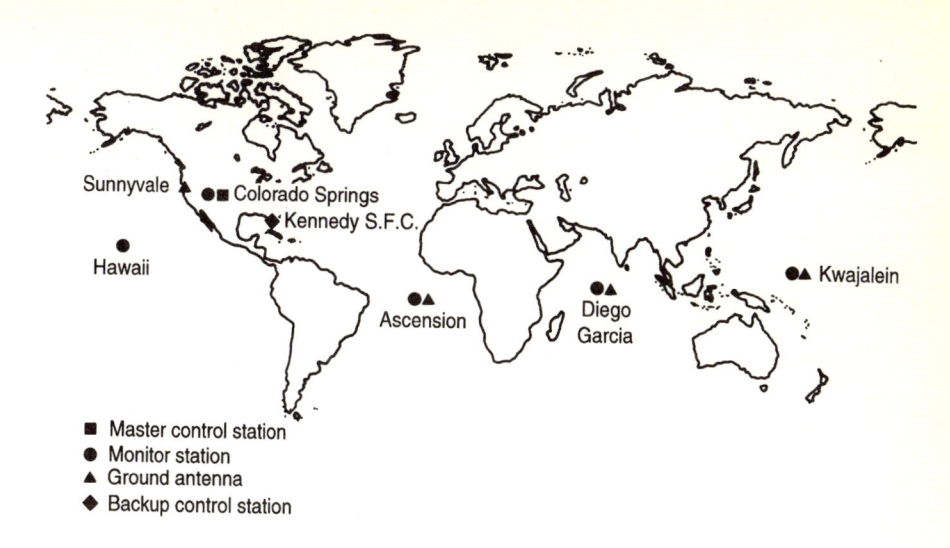

Fig 35 – The OCS

others requiring the maximum possible accuracy without needing instant real-time results.

Two of the regular tasks of the MCS are to manage satellite power systems during eclipses – when a satellite is behind the Earth and unable to see the Sun and has to run on its internal batteries – and occasional re-positioning to maintain satellites in their correct positions in the constellation.

Orbits and time offsets are calculated continuously so that up to four uploads per day to as many as four satellites simultaneously can be performed. System specifications require that the maximum errors in estimated satellite position and clock offset combined do not exceed six metres between uploads.

The ultimate aim is to make the system less and less dependent on regular uploads – obviously, it would be easier for an enemy to take the system out by destroying its master control station rather than by attempting to knock out all the satellites.

The MCS also has an important part to play during the launch of a new satellite. Up to the point where the satellite separates from its booster rocket, it is the responsibility of a ground launch crew, but after that the MCS takes over. Launches from Cape Canaveral mean that this point is achieved about thirty minutes into the launch, and contact is made through the monitor at Diego Garcia. The satellite at that point has been placed in a circular low-Earth orbit by its booster and has next to be kicked into a highly elliptical transfer orbit with its apogee at the height required for final orbit. Once stabilised in this orbit, the final motor burn takes place while at apogee and is calculated to produce a circular orbit at that height. When it arrives at this final orbit, its solar panels are deployed (it has been running on batteries until now), three-axis stabilisation achieved, and its navigation payload turned on.

That is not the end of the story. Another two weeks are needed before the satellite can be released for operational use. The quality of the radio-frequency signal must be checked and adjusted if necessary; the stability of the on-board clocks checked; the clocks slewed onto system time; and the vehicle parameters entered into the system Kalman filter. Only when the new satellite has maintained a steady performance for several days and all parameters are within tolerance is it released for use.

Uploads are made into each satellite daily and although assembled and checked at the MCS can be made through any of the OCS monitor stations.

Constellation performance information is sent from the MCS to the US Coastguard's Navigation Information Service, the designated point of information for all civil users. Scheduled and real-time satellite outages are reported in a standard form called 'Notice Advisory to Navstar Users', or NANU. Advance information of satellite shut-down for maintenance purposes is given whenever possible. These occur fairly regularly for station-keeping maneouvres and for on-board frequency standard maintenance (the caesium standards require ion-cleansing every eighteen months).

It is intended during 1995 to introduce new software at the MCS that will permit faster response to satellite problems.

5. GPS Errors and Selective Availability

The major error sources of GPS are:

1. Ionospheric Propagation Delay
2. Tropospheric Propagation Delay
3. Multipath
4. Satellite Ephemeris
5. Satellite Clock Drift
6. Receiver Clock Drift
7. Receiver Noise
8. C/A Code Selective Availability

Errors 1, 2 and 3 have been dealt with in Chapter 2 and are not significantly different for GPS from any other satellite navigation system.

5.1. Satellite Ephemeris

Even without SA (see below) there may still be small residual errors in the transmitted ephemeris. The specification for GPS is that they will be held at not worse than 3.7 metres (1-sigma), but there may be times when they are larger. Provided they are not too large (30 metres or more) they can be almost entirely compensated by differential techniques over moderate ranges, but there will be a gradual loss of accuracy as distance of user from monitor increases.

5.2. Satellite Clock Drift

This does not refer to absolute clock offset, which is given in the satellite message, but to drift since it was measured. The time of applicability (TOA) – the time when offset was measured – is given in the message as is also an estimated clock drift rate. The receiver computation applies the drift rate over the elapsed time since offset measurement and applies a correction. Since it is only a forecast there may be small errors.

5.3. Receiver Clock Drift

Again, this is not receiver clock offset, which is part of the measured pseudo-range and is taken out in the course of the normal receiver computations, but jitter or drift between measurements. Modern receivers employing parallel tracking have reduced this error to very small values.

5.4. Receiver Noise

Since GPS was designed in the 1970s there have been considerable improvements in the design of low-noise RF amplifiers. Reduction of the RF noise they generate has had the effect of improving signal/noise ratio quite considerably, meaning better accuracy. The general design of the rest of GPS receivers has also improved greatly and further assists this. The net effect is that modern receivers can now produce good results with much weaker signals than was the case in the 1970s, improving performance when the direct path to the satellite is obstructed. Unfortunately this does not mean that they are more resistant to interference. In modern receivers the contribution to fix error made by internal noise is a random effect of magnitude about three metres.

5.5. Selective Availability

The US Dept. of Defense considers that to permit the unfettered use worldwide of the highest accuracy levels of which GPS is capable would be unwise, and they therefore encode the P-code signal in such a way that civil users cannot use it, and then artificially induce errors into the C/A signal. The aim is to limit the achievable accuracy from GPS for civil users to about 100 metres (2-sigma). This figure was chosen to permit the use of GPS for as many civil applications as possible without providing non-US military users with a capability they could not already achieve by other methods. The full accuracy of the C/A code signal without SA would be in the region of forty metres, and its degradation to 100 metres is achieved by jittering the satellite timing slightly in an unpredictable fashion and by drifting the broadcast ephemeris. Military users have equipment that eliminate these effects.

The effect of SA is that indicated position, if the user is stationary, wanders around true position in an unpredictable fashion. The magnitude depends somewhat on the actual receiver, and the purpose for which it is designed. Some survey-quality receivers exhibit a smaller wander than others designed for aviation use. Wander of about one-and-a-half times the magnitude of horizontal wander also occurs in the vertical plane.

The US Federal Radionavigation Plan (1992) states that it is the aim of the DoD to keep overall GPS fix errors to not more than 100m horizontally (2drms) and 156 metres (2-sigma) vertically, world-wide. It also states that the horizontal error will not exceed 300 metres at the 99.99% probability level. This does not mean that errors will always be this large but it must be assumed in critical applications that they *may* be of this magnitude. A plot of SA-induced fix wander taken with a typical receiver is given at Fig 43.

The receiver used was a NOVATEL 951R, tracking eight satellites in parallel. Receivers that do not use parallel tracking or do not track all satellites in view may give different results.

The only way of reducing this error is to use a differential system (see next chapter), but for all normal en-route navigation this is not necessary.

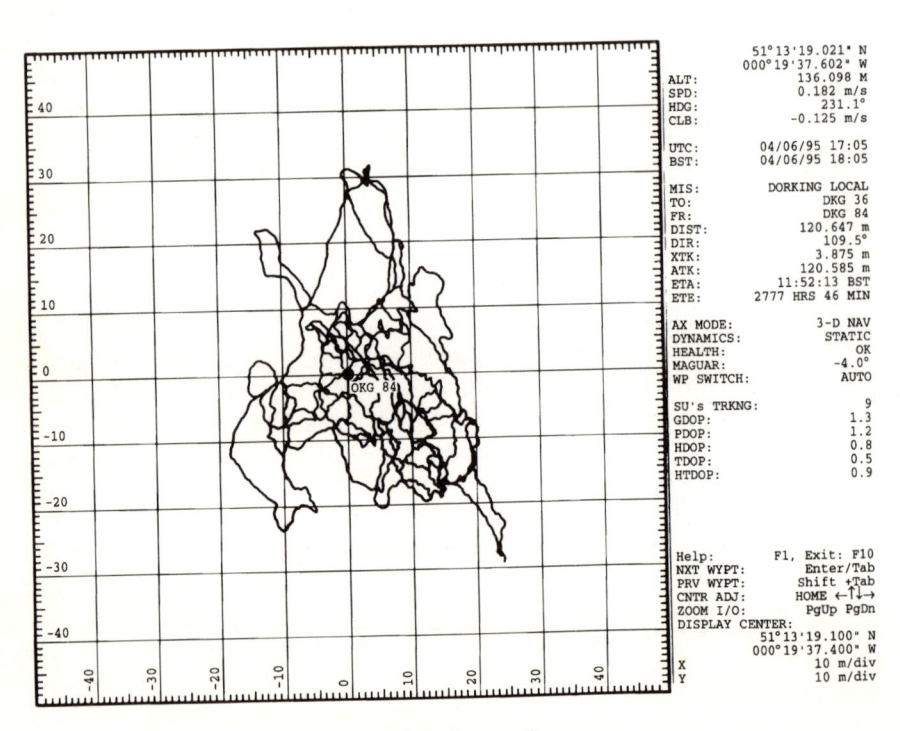

Fig 36 – SA wander

6. Differential GPS

A standard method of improving the accuracy of any navigational aid is to set up a monitor at a site whose position is already known very accurately and measure apparent errors. The navigator, who gets this information by radio link, can then use it to improve his own data. Care must be taken that system errors correlate over the distance covered by the data link; if errors are truly random then they will not and there is no point attempting correction in this way. Most radio-navigation systems have errors that do in fact correlate over reasonable distances and GPS is no exception. Differential GPS techniques are discussed in detail in Chapter 5.

7. GPS Integrity

This is dealt with in Chapter 8.

8. Reliability

The Federal Aviation Agency has its own monitoring system which checks the performance of the civil SPS signal as received off-air. It looks for continuity of fixing performance and the accuracy provided, using three monitors across the USA.

It classifies as a 'minor failure' any occasion on which a satellite exceeds a range error of thirty metres, and as 'major' if it exceeds 150 metres. 150 metres was chosen because it can translate into a fix error of 600 metres (assuming the maximum permissible PDOP of 4), the maximum allowable error in the system when used for en-route navigation.

Over the period from early 1993 to mid-1994 it found that there was one 'major' and one 'minor' failure of a Block II satellite, lasting a few minutes in each case. The monitoring system does not concern itself with the origin of each problem and in both cases the GPS OCS attributed it to errors at the OCS, which had still not been declared operational at the time.

The overall service reliability was found to be 99.97%, and the operational horizontal accuracy was measured at sixty-five metres 95% of the time, with a daily maximum error of 138 metres on average. Vertical error was 99 metres 95% and 207 metres average daily maximum. These figures are well within the SPS specification – see the Federal Radionavigation Plan Statement, Appendix 8.

Chapter 4

Aerials and Receivers for GPS

It may be a truism to say that without a good receiver all the effort put into the rest of a satnav system is wasted, but it is surprising how often little attention is paid to the receiver and its aerial. Correct selection of a receiver-type appropriate for the job in hand and proper installation of the aerial makes all the difference between an excellent system and one that only barely works.

1. The Aerial

The aerial the satellite uses for transmitting is designed to provide a constant level of signal all over at the Earth's surface no matter at what elevation above the horizon it may be. Fig 37 is the specification for the power received from a GPS satellite at between 5° and 90° elevation above the horizon (L1 signal).

There is only a very small variation – about 2 dB – between these extremes, which in practice is hardly detectable. Even down to 0° there is only another ½ dB fall-off so a good receiver aerial on a good site should enable the receiver to track satellites right down to the horizon. (This ignores a software feature of some receivers that forces them to drop tracking when the satellite is below 5° even if the signal is good). Therefore, if your receiver is unable to track anything below 30° then there is something wrong either with the receiver or the way its aerial is installed!

The satellite aerial designer has a fairly straightforward job to do. From a GPS satellite, the Earth subtends an angle of only 14° so the satellite aerial only has to produce constant power over this angle plus a bit to allow for slight errors in satellite pointing, which will not be worse than a degree or two. The total beam width of perhaps 20° permits an aerial design having a fair amount of power gain to make up for the usual fairly low-powered satellite transmitter.

The receiver aerial designer, on the other hand, has a much greater problem. He has to design an aerial that will pick up satellites in any direction at any elevation, in other words, with a hemispherical pick-up pattern.

A horizontal all-round pick-up pattern is not difficult to achieve; the problem is in the vertical plane where allowance must be made for the aircraft banking, climbing and diving – it would be most disconcerting to lose

Aerials and Receivers for GPS

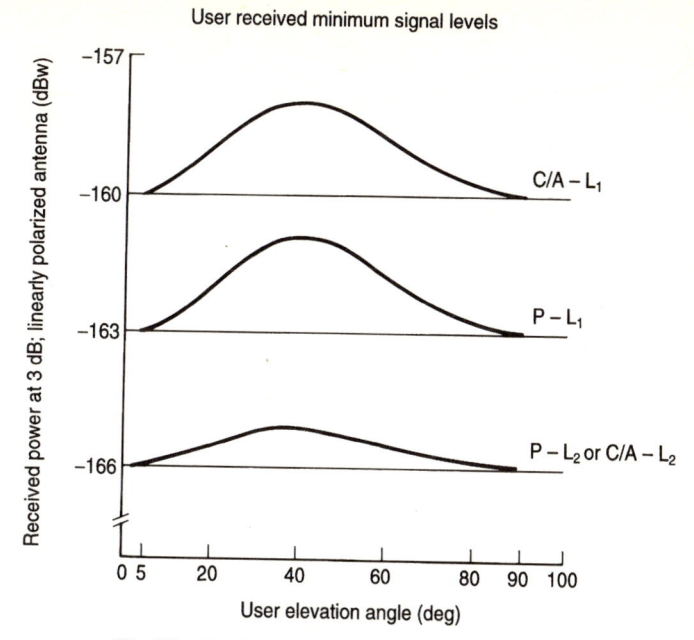

User received minimum signal levels

[Graph showing Received power at 3 dB; linearly polarized antenna (dBw) vs User elevation angle (deg), with curves labeled C/A – L$_1$, P – L$_1$, and P – L$_2$ or C/A – L$_2$]

Fig 37 – Satellite Power v. Elevation Angle (L1)

satellites the moment any departure from straight and level flight was made. It is impossible to make an aerial with a completely spherical pick-up pattern; if nothing else, it has to be mounted on the fuselage which would get in the way – remember these are line-of-sight signals. The usual compromise is to design an aerial that *in combination with the fuselage skin* has a basically hemispherical pattern extended below the horizon by about 20° or 30° all round. This allows for aircraft movement up to those limits in any direction. (Fig 38)

Two points to note: (a) the aerial only has this pattern when properly mounted in the approved position, and (b) most multipath (reflected) signals come from below so the aerial cannot discriminate against them.

A further limitation on aircraft aerials is that they have to be aerodynamic and introduce as little additional drag as possible. This eliminates some of the best designs used for non-aviation purposes such as the quadrifilar helix and means in practice that only 'patch' types can be used. They can either be truly flat or contained in a small dome but because they are so small they have to be very carefully tuned in order to be efficient. They are easily de-tuned if they are painted, allowed to get oily or dirty or placed near metal objects above the aircraft's skin. However, when properly installed they provide superb performance considering their size.

These aerials also contain a built-in amplifier because the satellite signal is

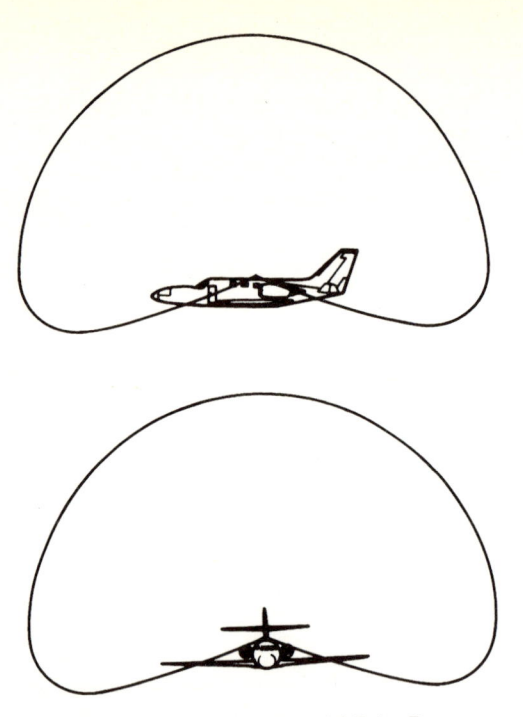

The Air Pilot's Guide to Satellite Positioning Systems

Fig 38 – Practical Aircraft Aerial Polar Pattern

so tiny that it would be lost down the connecting cable to the receiver unless it were only a few feet long. Although it ensures an adequate signal for the receiver it also introduces a few problems of its own. The GPS signal is a wide band signal (2 MHz) and it is impossible to design the amplifier so that it amplifies only the GPS signal and nothing else outside this bandwidth, as would be ideal. A nearby transmitter – the aircraft VHF communications transmitter, for example – might operate nowhere near the GPS frequency but may be so close it simply overwhelms the GPS amplifier. On a large aircraft the simple way round this is to make sure the two aerials are as well separated as possible, but it might not be possible on a small one. A much worse case is if the aircraft carries an Inmarsat system which transmits very close to the GPS frequencies, but only large aircraft might have one and they have space to separate them. There is another problem in that some VHF comms transmitters produce significant amounts of radio energy at harmonics of the basic frequency they are using which fall into the GPS band. Thus, the 13th harmonic of anything between 121.05 and 121.35 or the 12th harmonic of 131.125 to 131.5 falls across GPS and cannot be eliminated at the GPS set. They can be radiated from VHF transmitters complying in every way with specification and do not indicate a faulty transmitter, the problem arising from sheer proximity. They are often very weak and separating the aerials by only

Aerials and Receivers for GPS

another foot or two can make all the difference. Failing that, a special filter on the transmitter might be needed.

If the site for the GPS aerial is properly checked and a good installation made there will probably not be any problems of this type but it should be obvious that aerials stuck on the inside of cockpit canopies and so on cannot possibly perform properly and should be avoided at all costs.

2. Receivers

The aerial delivers to the receiver a very complex composite signal made up of signals from up to eleven satellites, all with different Doppler-shifts and timing delays, although all basically on the same radio frequency. The receiver must resolve this into individual satellite signals, make ranging measurements, decode their data and produce the required navigational information.

2.1 Finding the Satellites

The first process the receiver must go through is to find a satellite signal but unlike a normal radio receiver, it cannot even 'see' a signal until it has carried out basic PRN decoding. When it is switched on from cold for the first time it has no idea of where it is; which satellites are available, or whether its internal time base is correct. It knows what the PRN codes are for all the twenty-four satellites in the system (actually it stores thirty-two PRN codes) because they are part of its program, but it does not know which ones to use. If it is given no other information it has no option but to go into a general search mode to see which of the twenty-four are visible and this is often called the 'search-the-sky' mode.

In this mode, it takes all twenty-four possible PRN codes one by one and jumps each one by small increments across the entire spectrum where satellite signals might appear, waiting a short time after each jump to see if any RF energy is detected, which would indicate that a satellite signal with that code is present (see Appendix 7 for more details). It is a matter of luck how many PRN codes it has to go through before it finds one that matches a satellite signal and the process may take some time. Even when it finds one it still has to repeat the procedure three more times before it can get a fix. However, while it is finding the next three it can decode from the first one (now using the data code) the orbital, timing and almanac data. The speed with which it can do this is limited by the rate at which the satellite sends data and can take up to twelve and a half minutes. This information enables it to refine its internal time-base ('clock') and having found all four satellites it should now be able to produce its first fix although it may not be all that accurate. Further tracking refines its data and fixes will get steadily more accurate for a few minutes, after which it should settle down to proper operation.

This truly 'cold' start situation hardly ever occurs in practice because all receivers make provision for the operator to manually enter an initial approximate position and time, and, in some receivers, which satellites to look for. The 'time-to-first-fix' (TTFF) quoted by manufacturers always assumes that the operator has done this and all receivers store basic satellite almanacs, last-position and time information while switched off. When switched on again the receiver uses it to calculate which satellites are in view and their timing offsets and as a result the TTFF (now strictly speaking the 'warm-start' TTFF) may be reduced to a minute or two. However, if it has been moved significantly, perhaps through not being switched on for a leg of a flight, then the stored position will be wrong and correcting it will increase the TTFF again. Another factor that may increase TTFF is if it is switched on while in flight and it has to compute the Doppler-shift due to aircraft speed as well as its changed position.

2.2 Parallel and Serial Tracking

The strategy subsequently used to track satellites is subject to considerable variation in different receivers. Although there may be up to eleven satellites in view only four are needed for a fix so at one extreme receivers might track all eleven and thereby make sure that the optimum four are included, or it might simply pick the best four. If it tracks everything in view then while there is no need for complicated satellite selection algorithms it must include at least 11 tracking channels. Fig 39 shows a GPS receiver tracking ten satellites simultaneously.

If it tracks only the minimum four then it must have a good satellite selection system. There are all sorts of in-between stages between these two extremes, but the minimum number of channels is in practice usually five or six. This is because to keep DOP low new satellites must be brought in more frequently than might be necessary just because they get too low and the extra channel is used to find the next new satellite and feed it into the fix computation at the appropriate moment. The channel carrying the old replaced satellite is then released to go and search for another new satellite so as to ensure an uninterrupted flow of fixing information. Receivers these days are really special-purpose computers and it is old-fashioned to think of 'receivers' and 'computers' separately, but in the days when this terminology meant something, there was a great deal of discussion of the different merits of parallel and serial tracking and processing. 'Parallel' receivers had a separate channel for each satellite and so could track all of them simultaneously and continuously, while 'serial' sets had a single channel that was switched rapidly between satellites. There were even two types of switching – fast and slow. Fast switching sets were known as multiplex sets while slow switchers were called sequencers. Tracking in parallel had the advantages that it was continuous, smoother and could follow weaker signals than serial, while serial, because of its single channel, permitted a lower-cost set. As software tech-

Aerials and Receivers for GPS

NovAtel GPSCard™

```
┌─────────── Position ──────────────┐  ┌─────── Receiver ──────────────┐
│ LATITUDE:   51°13'17.776" N   σ 21.911 m │ MODE: 3-D NAV    HEALTH:  OK │
│ LONGITUDE: 000°19'37.696" W   σ  4.098 m │ DYNAMICS: STATIC IDLE: 24%   │
│ ALTITUDE:        112.823 m msl σ 26.213 m│ DATUM:                 WGS84 │
├─────── Time ──────┬──── Velocity ─────┤  │ SOLUTION:       SOLUTION OK  │
│ UTC:Thu 05/26/94 09:29 │ SPEED:  0.047 m/s │ ┌──────── Navigation ───────┤
│ BST:Thu 05/26/94 10:29 │ HEADING:  227.5 ° │ │ TO WAYPT:           DKG01 │
│ GPS:      0750.379753  │ CLIMB: -0.198 m/s │ │ FR WAYPT:           DKG02 │
├─────── Clock ─────┬──────── Dop ──────┤  │ DIST TO:        52.624 m     │
│ STATUS: SET, 1 PPS ADJ │ GDOP: 1.6  PDOP: 1.4 │ TRK DIST:      46.007 m  │
│ OFFSET: -5.1294E-02 us │ HDOP: 0.9  TDOP: 0.7 │ XTRK:         -24.048 m  │
├────────────────────── Channels ─────────────────────────────────────────┤
│ CHAN:    1     2     3     4     5     6     7     8     9    10        │
│ PRN:    20    12     6     4     5     1    14    24     9    25        │
│ STATE: Lock  Lock  Lock  Lock  Lock  Lock  Lock  Lock  Lock  Lock       │
│ dBHz:  48.1  45.0  41.3  35.8  48.8  47.3  37.9  31.2  45.5  42.3       │
│ AZ:    256°   91°  191°   66°   64°  281°  357°  101°  122°  299        │
│ EL:     62°   45°   20°   17°   76°   46°    1°   15°   39°   10        │
│ DPPLR: 1646 -2304  3669  1015  -817   486  1595  2564 -2749  3549       │
│ RESID: -0.24 0.83  0.12 -22.97 -25.33 3.23 -19.00 -11.51 -23.98 31.94   │
├─────────────────────────────────────────────────────────────────────────┤
│ Press F1 for help, F2 to toggle RESID/LOCK, F10 to exit                 │
└─────────────────────────────────────────────────────────────────────────┘
```

Fig 39 – 10 Satellites Tracked in Parallel

niques have improved, the distinction between them has become blurred and many different combinations are being used in modern sets. Naturally, each manufacturer claims some unique advantage for his particular method! Manufacturers sometimes do not state specifically which type their set is but it is possible to spot the difference by noting whether the number of channels in the set matches the maximum number of satellites that can be tracked simultaneously.

These technical differences are sometimes important when specialised work such as highly accurate surveying is to be done but for general aviation applications the small differences in performance are often swamped by other errors caused by aircraft speed and accelerations.

2.3 Velocity Measurement

Once at least four satellites are acquired, locked and being tracked, the Doppler offset due to a combination of aircraft and satellite movement can be measured and used as a direct indication of vehicle velocity. The velocity (heading/speed) readout presented on most GPS receivers is almost always calculated from Doppler-shift in this way because it is more accurate than calculating it from change of position over time in the traditional way.

2.4 Carrier Phase (Cycle) Tracking

Some receivers attempt to use phase measurements in addition to code to improve accuracy. One cycle of the L1 RF carrier has a wavelength of only

The Air Pilot's Guide to Satellite Positioning Systems

about 19 cms so accuracy can theoretically be improved to this level or better. It is not of much use for ordinary navigation but if the receiver is to be used for approach and landing then it becomes more important. However, there are problems doing this reliably at aircraft speeds and although feasible for static applications like surveying there is still considerable development to be done before it can be used for aviation. One of the basic problems is that if phase-tracking fails momentarily, tracking can slip off one cycle on to another and there is no easy way of correcting it. Although a one-cycle slip introduces only a 19-cm ranging error, the tendency is to slip a few hundred cycles each time before tracking is re-established. Even at only 150 kts the aircraft might be travelling through 406 cycles every second. The situation is similar to that of the lane/cycle ambiguity problem in Decca, Loran and Omega, but without their built-in ambiguity resolution methods. Eventually, new receiver design techniques may overcome this problem. At least one manufacturer (at the time of writing) now claims that the new methods of very accurate code-tracking ('narrow-banding') makes the fundamental code-tracking, that every receiver must do, so accurate that it can resolve this 19-cm ambiguity without external assistance.

Other methods in experimental use include recovering the carrier from the L2 frequency (this can be done without having to decode the P-code itself) and forming a new, lower, frequency from L1 and L2 that provides wider 'lanes' of about 80 cms wavelength that can be used very like the lane identification signal in Decca Navigator. One problem with this scheme is that considerable signal power is lost recovering the carrier which increases the likelihood of slipping these wider 'lanes'.

2.5 DGPS Inputs

Many aviation GPS receivers now incorporate the ability to use differential corrections and are stated to be 'Differential-ready'. Some caution is advisable. Although the current standard for differential correction transmissions is the marine RTCM SC-104 format, (see Chapter 5) there is now a new standard specifically designed for aircraft which will be much better than SC-104, which after all was only designed for slow-moving vehicles (see Chapter 8 for details). However, local DGPS systems, perhaps installed specially for a particular airfield and designed only for GA use, may be able to use SC-104 and find it perfectly satisfactory.

If a DGPS data-link and data-link receiver are installed that were not specifically designed for the GPS receiver, then even assuming they are electrically compatible (can talk to each other) it is worth checking that the receiver can handle the volume of data it will get. A five or six-channel GPS receiver presented with correction data for eleven or twelve satellites every three or four seconds might get a severe case of electronic indigestion!

A full discussion of differential systems is given in Chapter 5.

2.6 Data Outputs

The standard alpha-numeric presentation on an aviation receiver may not be sufficient for some users and provision is made on some sets for outputs to ancillary display and recording devices, often simple map displays. A data signal output is quite complex and a large number of parameters must be specified apart from the actual data to be sent. The electrical specification, for instance, must state the voltage levels, what they represent, and how they are to be activated. The 'ordinary' RS-232 data transfer specification beloved of computer manufacturers is open to widely differing interpretations and implementations in spite of apparently being closely defined, as anyone who has ever played with computers will know! And RS-232 only tells you what the electrical characteristics are – someone still has to define what data is to be sent, its format, and how it is to be dealt with.

To get over this, several 'protocols' have appeared and amongst them are the marine-oriented signal and data protocols known as NMEA (National Marine Electronic Association) Standards that have been adopted by many manufacturers of marine equipment. Much avionics GPS equipment is derived from or similar to marine equipment so these standards have also appeared in aviation receivers. The relevant standards are NMEA 0181/0182/0183/0185. They are only likely to be seen on sets destined for the general aviation market because sets for commercial aircraft must meet different aviation-approved data-bus standards set by EUROCAE in Europe and RTCA in the USA.

An NMEA output capability on a GPS receiver means that it can be coupled without further ado to any device capable of accepting an NMEA input and there are some add-on display devices on the aviation market of this type. But it does not mean that the add-on will accept *all* the data a receiver can give it and it is wise to check which parts of the receiver data output it will use. Another point worth checking is that as the NMEA data transfer rates and types were designed for rather low-speed marine applications they may be found to be too slow for the higher speeds and accelerations of an aircraft.

Non-NMEA data outputs may be of standard computer type (RS-232; IEEE-488, etc) which although specifying electrical characteristics say nothing about the actual data, speed of transfer and volume which are left to the designer of the add-on equipment to decide. While this gives him considerable freedom it often results in a data transfer system specific to only one type of device.

If the intention is to obtain a GPS receiver for use with an ancillary device it is as well to be very careful about whether one will drive the other or not, and even if it does, whether the data type and rate is sufficient for an aircraft.

Particular care should be exercised over map displays. It may sound attractive to have a map that can be expanded electronically to any desired scale, but just because it is bigger on the screen it is not necessarily more accurate. PC users will be familiar with graphics pictures that are claimed 'can be blown up to any size' when all that happens in practice is a better view is obtained of the pixels making up the picture rather than more detail of the picture itself! Electronic maps have sometimes been scanned-in directly from paper maps carrying information that was originally never surveyed, or perhaps not even drawn, to any great accuracy.

Chapter 5

Differential Satnav

Introduction

Differential Satnav is a method of improving the accuracy of satnav in real-time by measuring errors at a known site and transmitting them to a remote user who then applies them to his own measurements to obtain greater accuracy (Fig 40).

The basic principle is not particularly new, having been used for many years in other radio-positioning systems – translocation with Transit is an example – but what is new is the use of a radio data-link to the user. In earlier systems errors did not correlate over any very wide area and it was not worth sending them over a radio link. In satnav they correlate over hundreds and perhaps thousands of miles but vary rather rapidly so the use of a radio link is important to ensure the user receives correction information within seconds of it being measured.

Although the general principles are applicable to any satnav system the main use of the technique so far has been with GPS and for this reason the rest of this chapter refers to differential GPS (DGPS). One of the greatest advantages when used with GPS is that it removes the errors due to selective availability (at least while SA complies with the US Federal Radionavigation Plan specification). DGPS would certainly not lose its value if SA were removed but it would become of lesser importance. It is all a matter of what accuracy is needed – a few hundreds of metres and both SA and DGPS are irrelevant; a few tens of metres and DGPS is nice to have; a few metres and it becomes essential.

The way in which errors are measured at the monitor station(s) and corrections derived are pretty much the same in every DGPS system; the real differences come in the data link and how it works.

1. Advantages of DGPS

1.1 Accuracy Improvement
With the correct data-link, accuracy can be improved to the region of 2–3m whether or not SA is in use; over a distance of 1,000 kms or more from a

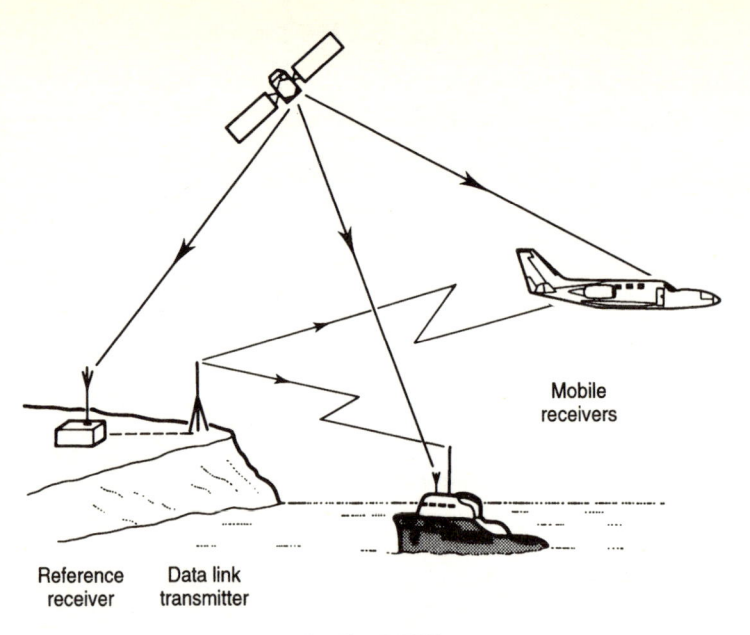

The Air Pilot's Guide to Satellite Positioning Systems

Fig 40 – DGPS

monitor station; on a twenty four-hour basis and unaffected by weather. If short ranges (up to 50 kms) are all that are needed, (for instance, aircraft landing), even greater accuracies can be obtained, perhaps to within 1–2 metres.

1.2. Integrity Checking
Since DGPS, in order to ensure its accuracy improvement, has to continuously monitor all aspects of satellite performance, any degradation will be observed immediately it happens and can be signalled to the user. This independent check reacts much faster than the standard GPS health word in the satellite message and can be made fast enough to satisfy integrity requirements for precision approach. A DGPS integrity system can also supply sufficiently detailed data for specialist users to decide whether certain problems affect them or can be be ignored. Integrity is a highly important factor for aviators and therefore it is treated separately in more detail in Chapter 8.

1.3. Permanent Recording of GPS Data and Performance
It is common practice for surveyors and other high-precision users to refine their received data by post-processing using records of the transmitted real-time data. For this to be done properly, not only the received data but also the

Differential Satnav

Table 8 – DGPS Accuracies Compared With Raw GPS.

C/A code only, SA applied (the normal situation):			
GPS basic	100 metres;	DGPS	3 metres
C/A code only, no SA:			
GPS basic	25 metres;	DGPS	1.5 metres
Both C/A and P codes, with or without SA (military only):			
GPS basic	12 metres;	DGPS	1 metre

transmitter records must be available. In the case of commercially-operated non-satnav positioning systems access to full transmitter performance records is part of the service offered, but detailed GPS system performance records are classified and not available to civil users. DGPS can provide this permanent record of GPS availability and accuracy as a by-product of its main activity.

2. DGPS System Components

2.1. Monitors

The purpose of these stations is to calculate corrections based on prior knowledge of their position. The very accurate receivers they use, and the fact that they are static, enable pseudo-ranges to be measured to centimetres so their position must also be known to the same accuracy. It is not always very easy to determine position that accurately, even for a fixed-site land station. Plotting its location on a large-scale plan and reading off latitude and longitude is definitely not good enough! Usually, GPS itself is used, which may seem a bit 'boot-strappish', but considering that some hundreds of thousands of 'fixes' can be taken in a few days and averaged it is not really so silly. Some special-purpose monitor receivers can do it themselves and need no more than to be placed on site and run for several hours in 'survey' mode. They calculate a mean position and indicate what accuracy this position has reached. The survey run can be stopped when position has been determined to whatever accuracy is deemed necessary, and this is then used as the reference position. This eliminates any need for accurate plotting on charts, datum conversions, and so on.

Having measured its own position, the monitor can revert to standard differential operation and start sending corrections. In early DGPS systems when there were very few satellites available, it was sufficient for the monitor to simply compare GPS fixes with its known position and transmit a geographical correction such as 'Move your fix 83 metres to the southwest' or whatever. When more satellites came along this method became

The Air Pilot's Guide to Satellite Positioning Systems

unsafe because there was no guarantee that the remote receiver was using the same set of satellites as the monitor. This is important because of the geometric (DOP) effects which are built into a geographical correction. If, for instance, the monitor used a set of satellites producing a DOP of 10 it might measure a fix offset of 60 metres and send that as a correction, whereas a user might be using another set giving a DOP of only 5 and producing a real fix error of only 30 metres. Applying the 60 metre correction to his 30 metres would, at best, do nothing to reduce the error and at worst could increase fix error considerably. Strictly speaking, if both receivers were absolutely identical; used the same algorithm for selecting satellites, and were close together then they would indeed select the same set, but in practice this does not often happen. It is anyway impossible to ensure that all users of a particular DGPS system will use only one type of receiver (although it might be done on a limited scale in a commercial system). Even then, receivers choose their four satellites on the basis of geometry and elevation which depend on user position, and within a 1,000 km range it is possible for users to be using many different combinations of four satellites. Some may even be using every satellite in view – many receivers now have this capability.

For these reasons the geographical fix correction method has dropped out of use and the pseudo-range correction system is now universally used. After determining its own position, the monitor receiver switches into 'differential' mode, measures pseudo-ranges to all visible GPS satellites, performs health checks, and formats a correction message. Corrections are calculated by first working out what the pseudo-range should be, assuming the receiver and satellite positions are correct. The measured pseudo-range is compared with this and the correction is the difference between the two. This can be done comfortably enough even with a simple GPS receiver by the addition of a standard desk-top or lap-top PC, although many receivers now have DGPS monitor and remote capability built-in.

Monitor stations often have duplicated receivers and computers, data extraction and formatting equipment, time standards, modems (if the data is to be sent to other sites as well as over the radio link), and uninterruptable power supplies. They will probably be automatic and unattended, and may need to be fairly small and easily transportable if not permanent. Small local-area schemes, such as on airfields, need only a low-power VHF or UHF data-link transmitter to cover a few miles radius. In large schemes covering thousands of square miles, data transmission would probably be carried out from some centralised point and would carry correction data for several monitors simultaneously.

2.2. *Central Control Station*

These are only required for large and permanent wide-area systems using several monitors. They need high-speed computing that may be more than a

Differential Satnav

PC or two can cope with, and because of the large number of users depending on them they must perform very comprehensive error-checking. A CCS accepts the data from the system monitors, checks validity, does all the system computations and the final formatting of the message which it then sends to the data link transmitter(s). It will also have equipment similar to that its users have so that it can receive its own transmitted data back off-air. Besides comparing this with its original out-going data it uses it to correct a local GPS receiver in exactly the same way as a user would do, logging the remanent errors. If these exceed some predetermined limit it triggers its own warning system.

The CCS can be at any convenient location and does not have to be a monitor itself. This allows the location of monitors to be at the best possible sites for clear GPS reception which are often in rather remote locations. It also permits the computation of correction data by collating results from several monitors and adding detailed ionospheric corrections. This is the type of system being adopted for the Wide-Area Augmentation System (WAAS) proposed for aviation purposes (see Chapter 8).

2.3 The Data Link

This is probably the most important part of DGPS; on it depends whether the system will work at all. It is not as easy as simply picking a good radio channel and sending the differential data over it; the amount of data, its repetition rate and required reliability are all important factors in the selection of the correct type of data link.

2.3.1. The Data To Be Sent

Transmitting corrections for all visible satellites imposes a considerable computational and data link load, particularly when range-rate and phase corrections are also sent, and this has caused data rates to increase steadily from the original GPS-compatible 50 bps specified in SC-104. There is much more to be sent than merely the pseudo-range corrections, even if the system is dedicated only to differential data and not the more complex aviation differential/integrity messages. For the moment we will consider only basic differential data – the additional data needed for aviators is described in Chapter 8. Table 9 shows the different types of data message allowed in the marine RTCM SC-104 DGPS data format, widely used in the offshore industry and for marine purposes. Because it was established some years ago and until recently was the only available standard, it has been adopted for non-marine purposes as well as marine.

Table 9 – Message Types

Message Type No.	Current Status	Title
1	Fixed	Differential GPS Corrections
2	Fixed	Delta Differential GPS Corrections
3	Fixed	Reference Station Parameters
4	Tentative	Surveying
5	Tentative	Constellation Health
6	Fixed	Null Frame
7	Tentative	Beacon Almanacs
8	Tentative	Pseudolite Almanacs
9	Fixed	High Rate Differential GPS Corrections
10	Reserved	P-Code Differential Corrections (all)
11	Reserved	C/A-Code L1, L2 Delta Corrections
12	Reserved	Pseudolite Station Parameters
13	Tentative	Ground Transmitter Parameters
14	Reserved	Surveying Auxiliary Message
15	Reserved	Ionosphere (Troposphere) Message
16	Fixed	Special Message
17	Tentative	Ephemeris Almanac
18–59	—	Undefined
60–63	Reserved	Differential Loran-C Messages

Message Types

Type 1.
Contains pseudo-range corrections for all the satellites a monitor can see at a given time. In addition to the basic pseudo-range corrections, estimates of their accuracy; the rate at which they are changing; and the time of validity (Issue of Data – IoD) are included.

Type 2.
Provide data to allow for the possibility that the monitor and the user receivers are not using the same data message from the satellite. This problem arises from the fact that when the satellite data is renewed by the uplink station it may not be downloaded by all receivers at the same time. Some receivers only look for new orbital data every half-hour so it is possible for the remote and monitor receivers to be calculating their pseudo-ranges using slightly different satellite positions.

Type 3.
Give the monitor station position in ECEF x,y,z co-ordinates. This is not essential and is often omitted.

Type 4.
For the use of high-accuracy survey receivers. Provides a differential phase correction. Only transmitted when centimetric accuracy is required.

Type 5.
The DGPS-originated system health message. It allows the monitor to perform its own health checks on a satellite including measurements of signal strength and permits the monitor to override the satellite's own health signal.

Of the other messages only the Type 12 is of interest to aviation users. It gives data concerning pseudolite stations that may be used at airfields for approach and landing purposes. Not all of these messages need be sent, particularly where aviation is concerned. A simple local-area system could dispense with almost all messages except 1 and 5.

2.3.2. Latency and Data Rate

An important measurement of performance of a DGPS system is latency. By this is meant the lag between measuring the correction and the receiver getting it. Apart from delays due to computation the digital data rate of the correction transmission must be fast enough to keep latency to a reasonable figure, but what exactly is 'reasonable' depends on the application. For en-route navigation a latency of 30 seconds may be acceptable but for precision approach a much shorter period must be achieved. For example, the US WAAS specification (Appendix 12) allows no more than a 5.2 second latency which includes all transmission and computation delays. Where the data link is a geostationary satellite the signal propagation delay alone is of the order of a quarter of a second.

If the GPS satellite errors were changing only very slowly, then the data rate itself might only need to be quite slow, and this is true of what might be called the 'natural' errors. However, SA changes quite rapidly, and needs an update rate of not more than about thirty seconds if its contributory error is to be kept below five metres. The total requirement is to update data for up to eleven satellites, with virtually no errors attributable to the data stream itself, at a rate no slower than about once every thirty seconds. In some DGPS systems, there will be several monitors and data for all of them must be transmitted over the data link so the user can choose the most appropriate and also have some fall-back in case of one failing.

SC-104 specifies a data rate of 50 bps, and a digital data word structure similar to that of GPS itself. The original idea was that it could be decoded by the same processor and software as used in standard GPS receivers, all that needed changing being a few software routines. In practice, this has not happened, most users using a separate PC. This rather slow data rate has meant that although SC-104 is perfectly usable for small single-monitor systems it cannot be used for larger wide-area systems where there is much more data to be sent.

Accepting the need to renew everything at no more than thirty second intervals and perhaps as quickly as five seconds, the amount of data itself is important. The SC-104 recommendations allow a choice of messages depending on the use to which the corrections are to be put and for offshore survey a representative set of messages might be types 1,2,3,4,5,7 and 9 which for eleven satellites represents about 2,500 bits of data, taking 50 seconds to send at 50 bps. This is just not fast enough, particularly if it is decided to repeat all messages twice in case of difficult reception. At 50 bps a message repeated every 10 secs could only contain little more than the basic identifier and type 1 message for eight satellites.

2.3.3. Data Link Bit Error Rates

There is obviously little point using a DGPS data link that itself has low reliability. Ideally it would have an inherent error rate no worse than that of GPS data itself, about 1 in 10^6. Most digital radio transmission methods using anything other than a satellite link have a hard time meeting even a 1 in 10^4 level, representing a loss of no less than one message in thirty-three, (or about one every two minutes at 50 Bd if the RTCM format is used). Using modern digital error recognition and detection techniques such as half-rate convolutional encoding and interleaving can help, but all such techniques derive their improvement mainly by transmitting more than the theoretical minimum number of bits needed to represent each digit and therefore the bit count soars for a modest improvement. A 50 Bd data rate becomes a 100 bps bit rate with half-rate coding and it is the actual bit rate that is important when designing the radio side of the data link although it is the data rate that is important for DGPS. Thus, comprehensive bit error detection and correction systems may themselves add substantially to the data rate and can only be realistically accommodated over fast links carried on satellite channels.

2.3.4. Transmitter Radio-frequency

It is economically important to match data-link range with the required correction area. Links based on line-of-sight propagation (VHF, UHF, microwave) unless transmitted from a satellite, are limited to under 100 km and so are usable only for local areas. The radio-frequency band of 300–500 kHz offers the possibility of greater range and is being used for marine DGPS. Data is transmitted either over special DGPS-only transmitters or via existing air and marine radiobeacons. These latter are severely limited in range, there being only five in the whole of Europe that are designed to cover as much as 150 km at night. Most aviation MF beacons are limited to far shorter ranges of 10 to 50 kms, even in daylight, worse than VHF (except perhaps for low-level flights). This band suffers an inescapable night-time sky wave problem and although range in daylight might be considerably better than at night, users of a DGPS system expect the data link to work as well as GPS itself – 24 hrs/day without interruption.

Even if range was to be discounted, bit transmission speeds at these frequencies are limited by technical characteristics of the transmitters, aerials and propagation to something in the region of 200 bps – which may mean an

Differential Satnav

actual data rate of less than 100 Bd, as we have already seen. Although there is not much technically to prevent such transmissions being used for aviation purposes if suitable equipment is available, their reliability is so low as to preclude serious consideration for landing purposes.

Higher in the radio-frequency spectrum, HF radio can cover thousands of kilometres but the problem there is that the optimum frequency changes three or four times a day and even when the best frequency is chosen the error rate is rather high. By using three frequencies in parallel, the error rate can be reduced to something acceptable and can provide wide-area coverage. A limitation is that, like MF, bit transmission rate at these frequencies is limited by propagational characteristics to 100 bps. This is too low for most aviation applications even if the need for switching frequencies frequently is acceptable.

From most technical viewpoints, satellite data links cannot be bettered – they can easily meet all range, reliability, and low error rate requirements and are capable of supporting virtually unlimited bit rates. It is for these reasons that satellite links are being specified for all proposed wide-area DGPS systems for aviation use. The two drawbacks are cost and the low signal level when used with non-directional aerials as used on aircraft. Both can be overcome if the system is destined to be used over a wide area by many users and the expense of a high-powered system can be justified. This, of course, is the case with WAAS and similar systems.

2.3.5. Pseudolites

These are a specialised type of microwave beacon used as a differential data link but also transmit GPS-lookalike signals. To eliminate the need for a special receiver they transmit on the same radio frequency as GPS and use the same type of modulation. In fact, as far as the GPS receiver is concerned they are simply GPS satellites on the ground and they are treated the same way except for their additional differential data. Located at an airport they act as local monitors and increase GPS accuracy so that aircraft can use GPS as a landing aid. Power is kept very low to avoid interfering with the real satellites that are much further away and have very weak signals compared with the pseudolite. Because of this very limited range (a few kilometres at most) they are unlikely to find much application elsewhere. See Chapter 8 for more details.

2.3.6. Use of Existing Positioning System Transmitters for the Data Link.

DGPS data can be transmitted over navigational transmitters and many navaids now carry it. The main reason has been to provide supplementary services to the main navigation aid itself, such as ambiguity determination. Systems that are being used in this way are the offshore positioning systems like Argo; Hyperfix; Pulse/8; Geoloc; Spot; Starfix; and others.

While useful where these systems already exist, none of them was designed primarily as a data carrier and such a function must be subordinate to their main purpose. Data rate and latency are usually poor and the data sent very limited.

For aviation purposes the use of systems like MLS, SSR and ACARS to carry DGPS signals has been proposed, but it seems one of the most likely methods is to use a number of channels in the existing VOR VHF allocation (112–118 MHz). If satnav comes into wide aviation, application this band will gradually empty as VORs are withdrawn and their frequencies will become available. A draft specification has been drawn up (Special Category 1 DGNSS Instrument Approach System – SCAT–1) and trials are under way. These channels will be allocated on a 25 kHz channel spacing basis and the data link will be digitally modulated at a bit rate of 31.5 kHz. Its technical characteristics are such that up to four runway ends per airfield can be covered with one channel.

3. DGPS Error Sources

3.1. Errors Affecting Only Civil Users

There are three sources of DGPS error that affect civil users (as distinct from military systems):

(a) GPS transmits two radio frequencies in order that the military user can eliminate ionospheric-refraction errors in real-time. Civil users are allowed to use only one frequency and cannot correct directly.

The characteristics of ionospheric refraction (Chapter 2) are such that although rather variable with time it correlates quite well over short and medium distances and thus meets a basic requirement for differential correction. In disturbed conditions (magnetic storms, etc) it may not correlate well and DGPS may not be able to correct properly, perhaps even, in extreme cases, making the situation worse. The smaller the area involved the better, from this viewpoint, and over very small areas like that of the average airfield and its surroundings it should work well whatever the ionospheric conditions. Large systems like WAAS will have monitor stations making ionospheric observations to enable the construction of a real-time ionospheric model from which corrections can be deduced and transmitted.

(b) The electronic characteristics of the spread-spectrum modulation applied to the C/A signal (1 MHz as against 10 MHz for the P-code) inherently provide less accurate measurements.

There is nothing much to be done about this – it is a basic technical limitation. The errors it causes are peculiar to each receiver, are totally random and cannot be corrected. However, they are considerably smaller than the 1 to 10 frequency ratio might suggest – see Appendix 7.

Differential Satnav

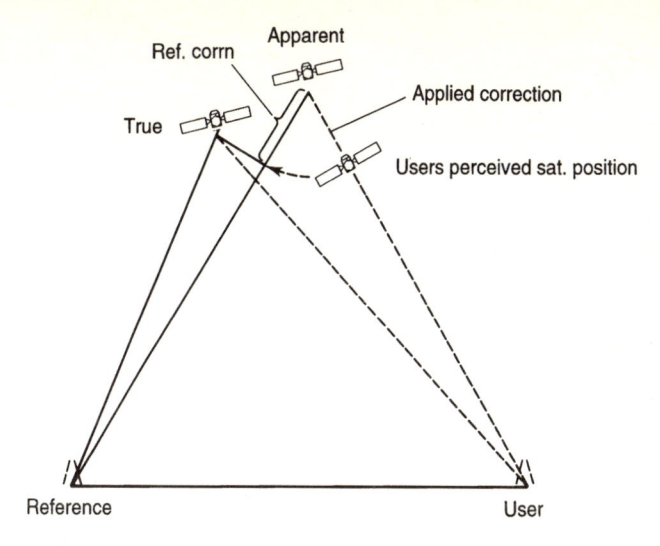

Fig 41 – DGPS Geometric Error

(c) The ephemeris and basic timing information provided to the civil user on the C/A channel are deliberately degraded by selective availability.

The effect of SA, as already seen, is to make position wander rather slowly in an unpredictable fashion around the true position. The rate at which it wanders is important because (i) it determines how fast corrections must be made, and (ii) if the rate of wander is varied very rapidly then the receiver itself may lose track. This, of course, could be used as a way of defeating DGPS systems if it became necessary. The shift in timing induced by SA remains the same over the entire area of coverage so it correlates exactly everywhere, but the other part of the wobble – an artifical variation in orbital parameters – does not. Its effect is to cause both the monitor and user-receivers to calculate the same slightly incorrect satellite position, which has geometric effects depending on where the receivers actually are. (Fig 41).

3.2. DGPS Error Sources Affecting All Users.

(a) Tropospheric refraction.

The same considerations apply as given in Chapter 2. It is very localised and can be eliminated at the receiver by the use of a suitable correction algorithm if the right meteorological information is available. DGPS cannot help over large areas where the atmospheric conditions may be very different but it can help over small areas of less than 50 kms or so and would certainly be useful over the area of an airfield.

If the DGPS system is to be used by high-flying aircraft then the monitor stations, being at low altitudes, must take out their own tropo bias before transmitting corrections. If they do not do this, the tropo bias they are experiencing and appropriate at low altitudes will be transferred to the aircraft which being high will not be experiencing the same effects.

(b) Multipath.

This, like receiver noise, is extremely local to each receiver and since it depends on what reflecting surfaces are nearby is not reproducible between receivers. DGPS cannot help in this situation. Its reduction is a matter of aerial design and siting.

3.3. Other DGPS Accuracy Considerations

3.3.1. Receiver Type

The use of different types of receiver for monitoring and receiving can cause problems because of differing internal design philosophies. What causes this is that receiver 'clock offset', nominally the difference between the receiver clock and satellite system time, always incorporates other small delays caused by the electronic design of the receiver. The type and operation of filters and where they are placed are important factors. For instance, in some receivers the pseudo-range measurement is available at a very early stage and has been subjected only to mild filtering, while in others it has passed through a complete Kalman filter. A general-purpose DGPS system has to recognise that it must cater for many different types of receiver and the monitor receiver must be carefully designed not to add biases peculiar to itself.

3.3.2. Differing Processing Algorithms

Even when all possible precautions have been taken to commonalise data, fix results from nominally similar receivers but made by different manufacturers have been known to vary slightly. This is sometimes found to be due to the use of different processing algorithms, one example being the correction for Earth rotation. No standard algorithm exists and the effect is differing receiver biases even when using identical data. Unfortunately it is often difficult to find out exactly which algorithms are being used because manufacturers are unwilling to disclose them. Of course, no manufacturer will admit that his particular method is not as good as a competitors'!

3.3.3. Monitor Clock Bias

Uncorrected monitor clock bias appears as a range bias and is transmitted as part of the correction. It does not normally matter because the user sees it as a fixed addition to all corrections and it will be taken out in the usual way. However, if he attempts to use one or two satellites without applying DGPS corrections to them, for instance if he can see satellites that the monitor cannot and therefore he has no corrections for them, then the additional bias appears

Differential Satnav

only on some satellites and a differential error will occur. What this means in practice is that no attempt should be made to stretch the range of a DGPS system beyond its designed range and that DGPS should always be operated on an all-in-view pseudo-range correction basis.

3.3.4. Time of Data Sets

It is usually assumed that both the monitor and user receivers are using the same down-loaded GPS data sets, (as identified by the IODE/IODC words), but the transmitted ephemeris is changed every hour and there is a chance that they may be using different data, since some receivers do not check for a new ephemeris this often. The standard data format for differential systems (SC-104) recognises this and makes provision for the broadcast of an additional set of corrections for use when the two do not match. A good DGPS receiver will of course provide for this.

4. Area of DGPS Coverage

Two factors affect the area of coverage – the area over which the monitors' corrections hold valid; and the area covered by the data link transmissions.

4.1. Area of Correction Validity

The area over which DGPS corrections apply when only a single monitor is used depends primarily on the size of the correlation areas for each source of error, as indicated above. The major error that is corrected by DGPS is SA and since the SA-induced pseudo-range error is the same no matter where the satellite is used, then the correction is also valid everywhere that satellite is used. This is not true for the SA-induced ephemeris error but the effect of that is much smaller.

The next largest error is ionospheric refraction and under undisturbed conditions this can also be constant over quite large areas provided a day/night boundary is not crossed. It should not be forgotten that if the ionospheric delay were the same for all satellites then no matter how large it was there would be no effect on final fix error. However, under disturbed conditions the ionosphere can vary considerably over small distances and this is no longer true. If the data from a number of monitor stations is reduced collectively, additional benefits accrue such as a new ephemeris which can be transmitted over the data link and used instead of the broadcast ephemeris. If each monitor also measures its local ionospheric density and in addition data from specialist ionospheric monitoring stations is incorporated then a very accurate picture of the ionosphere can be built up in real time and its effects on single-channel receivers estimated. The WAAS system proposes to broadcast iono corrections valid for every ten square kilometre block in its coverage area.

To quote some practical results, the major users of wide-area DGPS so far have been offshore surveyors, who use several systems using geostationary satellite data links. Their tests have shown that even at as much as 2,500 kms monitor/user separation, accuracy is maintained at 10 metres. For moderate accuracies in the 5–10 metre range a single monitor will easily allow an area of radius 500 km to be covered.

It is not usually the area of validity of the corrections that limits DGPS coverage, but the radio propagation characteristics of the data link itself.

4.2. Area of Data Link Cover

Apart from sheer radio range, the area covered by the data link depends upon how reliably the data must be received. If very high reliability is necessary, e.g. for aircraft landing, then a line-of-sight link will be needed, either from a satellite or from a local VHF transmitter, but for other less critical purposes lower reliability data links such as medium-frequency beacons can be used. To cover very large areas, when the data from several monitors is combined into a single correction message, requires the use of a data link with a similarly large coverage area. Additionally, if the major use of this wide-area system is for integrity purposes then the data link itself must have very high reliability. A satellite data link is really the only way of doing this, and even then to make sure the data link is not itself the weak link may require transmission of the integrity/differential message via two satellites simultaneously as is proposed for WAAS.

Chapter 6

The Global Navigation Satellite System (GLONASS)

The first test satellites for the Global Navigation Satellite System, (GLONASS) were launched in the former USSR in the mid-seventies as the forerunners of a system designed primarily for the use of the USSR's Armed Forces. In 1987 it was decided that civil users would be permitted access and at a meeting of the ICAO Special Committee on Future Air Navigation Systems (FANS) in 1988, GLONASS was offered charge-free to the world community. A similar offer was made to mariners at the 35th session of the IMO Subcommittee of Navigation Safety. Since then, several types of receivers for civil users have been designed in CIS countries and the USA, some models now being produced in quantity. Recent economic problems have adversely affected the development of GLONASS space and ground control segments, but it is still fully intended to put GLONASS into successful operation. The official target date for the full 24-satellite constellation is 1995, but a more realistic date is probably 1997 or 1998.

1. System Description

1.1. Basic Concept
As with GPS, GLONASS users determine position by measuring pseudoranges to four GLONASS satellites, whose positions are obtained from data transmitted by each satellite as part of its navigation message. User position is given related to the Russian SGS-85 datum and the satellite time shift is given relative to UTC (SU) (UTC as measured by Russian observatories).

1.2 Space Segment
The fully-deployed space segment will consist of 24 satellites in three orbital planes with eight satellites in each plane. These planes are spaced 120° apart from each other (Fig 42) and the satellites will be evenly distributed around each plane, that is, they will be 45° apart. The system therefore has the same number of satellites as GPS but in three instead of six planes.

These orbits are virtually circular with eccentricities less than 0.01, revolution periods of around 11 hr 15 min, altitudes of 19,100 km and inclinations of 64.8°. By using this inclination, precession of the perigee point is min-

The Air Pilot's Guide to Satellite Positioning Systems

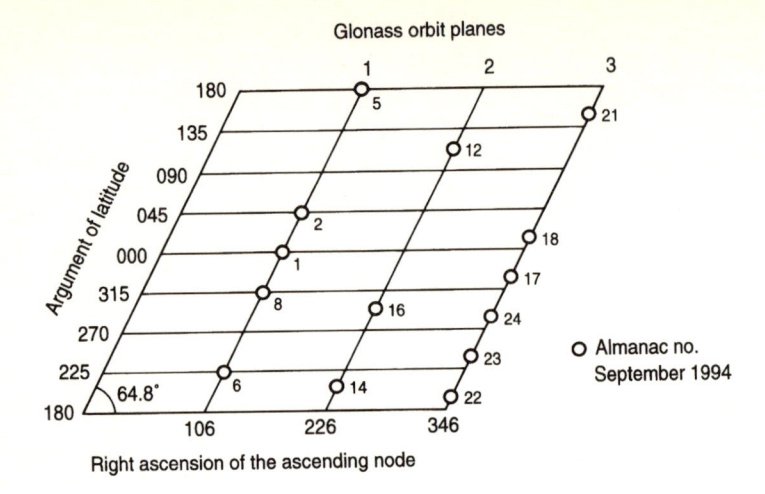

Fig 42 – GLONASS Orbital Constellation

imised (see Chapter 2) making predictions of future satellite position somewhat easier than for GPS. The lower orbit was chosen following an analysis of system accuracy which showed that more repeatable and consistent results in high latitudes would follow from this choice. In practice, even before the full constellation is in operation, GLONASS provides better coverage on average in European latitudes than does GPS. The satellites that were active in September 1994 are listed in Table 10.

The full system is designed so that at least eight satellites will be visible anywhere in the world at any one time. If the total number of satellites were to be reduced to twenty-one, only four satellites would be in sight (assuming a minimum acceptable elevation of 5°).

The launcher used for GLONASS (Proton booster) is powerful enough to launch three satellites simultaneously. To replenish the system when a satellite fails, either an orbital spare is used or a three-satellite unit is launched. If the latter is the case, one of satellites launched is used to replace the unhealthy one and the others become orbital spares. This method of launch requires some satellites to drift considerable distances around their orbital plane in order to reach their allocated planar positions. Occasionally this drift period has taken up to three months.

Each satellite transmits a navigation signal at two frequencies within the same general radio-frequency band as GPS but not on the GPS frequencies. As with GPS, the signals emitted at L2-band frequencies are encrypted and not authorised for international use, civil users being allowed only the L1-band signals. However, as with GPS the L2 signals can be used for measuring ionospheric delay without knowing the encryption code. The frequencies used in the L1-band are 1602.5626 MHz to 1615.5 MHz with a carrier sepa-

The Global Navigation Satellite System (GLONASS)

Table 10 – GLONASS Space Segment–Sep '94

Sat ID	Cosmos	GLN	CHN	ALM	Plane
1992– 5C	2179	55	23	1	1
1993– 10B	2235	60	5	2	1
1990–110C	2111	49	23	5	1
1993– 10C	2236	61	22	6	1
1992– 5B	2178	54	2	8	1
1994– 50A	2287	65	21	12	2
1994– 50B	2288	66	21	16	2
1994– 50C	2289	67	9	14	2
1994– 21A	2275	62	24	17	3
1994– 21C	2277	64	10	18	3
1992– 47C	2206	58	24	21	3
1991– 25B	2140	51	11	22	3
1994– 21B	2276	63	3	23	3
1992– 47A	2204	56	1	24	3Ï

14 Active GLONASS Satellites – September 1994

Fig 43 – GLONASS satellite

ration of 0.5625 MHz. The satellites are thus both identified and separated from each other by their carrier frequencies, which are determined using the following expressions:

$$f_{k1} = f_1 + k_1 \, sf$$
$$f_{k2} = f_2 + k_2 \, sf$$

where:

k=1,2,...24 (the operational frequency number)
f_1 = 1602 MHz; f_2 = 1246 MHz;
sf_1 = 0.5625 MHz; sf_2 = 0.4375 MHz

These frequencies are coherent and generated from a single on-board frequency standard. The relationship between the two operational frequency bands i.e. f2/f1, is 7/9, as required if they are to be used for ionospheric refraction calculation.

Unfortunately the frequency band of GLONASS satellite navigation signals partially overlaps a band allocated for radio astronomy and causes significant interference with some radiotelescopes. Because of this it is possible that GLONASS frequencies may be changed at some time in the future.

GLONASS satellites are launched three at a time from the Baikonur Space Centre. They are first launched into intermediate circular orbits 200 km in altitude and then transferred to an elliptical orbit 200 km at perigee by 19,100 km in apogee. When this has been stabilised and the dimensions verified a booster burn at apogee circularises the orbit. The specification for the maximum difference between the satellites' actual orbit and its design parameters is that they must not differ more than:

in period	±0.1s
in orbit inclination	±0.3°
in eccentricity	0.01
in latitude argument	±0.1°

A complete list of all GLONASS satellite launches is at Appendix 10.

1.4. Control Segment

Satellite position and timing information is generated by the monitoring/control segment and then uplinked to the satellite.

This segment comprises:
- system control center (SCC);
- five command/measurement stations (CMS);
- three stations with equipment for orbital monitoring (ENFM);
- two main measurement stations (MMS) containing central synchronizer (CS), phase monitoring system (PMS) and time-referencing equipment (TRE);
- four quantum optical stations (QOS).

These stations are located in the former USSR as follows:

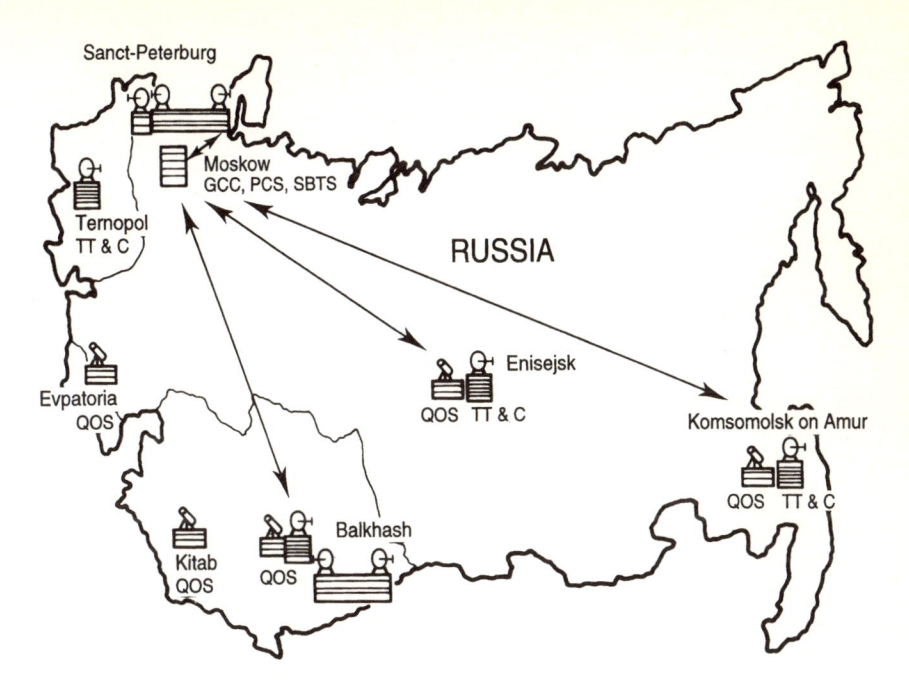

The Global Navigation Satellite System (GLONASS)

Fig 44 – Glonass OCS

Khmelnitsky area;
near the Balkhash Lake;
near Eniseysk;
near Komsomolsk-on-Amur.

The system control centre is located at the GLONASS technical headquarters near Moscow (Fig.44).

The GLONASS control segment functions in a way largely similar to that of GPS. The main system differences are the separate monitoring in GLONASS of ephemeris and frequency/time support and the use of two-way active measurement techniques for monitoring orbits and synchronizing satellite time standards.

The control and monitoring stations are located only on the territory of the former USSR and although this is a vast area it gives rise to a number of disadvantages:

* orbit/time measurements cannot be made continuously throughout the whole of each orbit resulting in accuracy degradation when computing orbits and frequency/time data;
* the inability to monitor each satellite continuously outside the territory of the former USSR results in degradation of GLONASS operational reliability as a whole and extension of the time interval needed to notify users of improper functioning.

1.4. System Parameters

The target objectives for determination of user's position and velocity are:

horizontal position	100 m (95%)
height	150 m (95%)
velocity	15 cm/s (95%)
Time offset against UTC (SU)	max. 1 ms.

These targets are of course very much the same as the performance of GPS's SA'd SPS, and are those given in the Russian system document submitted to ICAO.

Satellite orbital errors (co-ordinate and velocity vector components) must be no more than (1σ):

along-orbit	20 m and 0.5cm/s
cross-orbit	10 m and 0.1cm/s
vertical	5 m and 0.3cm/s

Time synchronization for all satellites is held within 20ns (1σ).

GLONASS does not employ special measures to decrease available accuracy to civil users, as does GPS (SA), and the practical accuracy obtainable by civil users is therefore somewhat higher. GLONASS tests have shown that practical navigation accuracy is at least twice its design value. Fig.45 provides accuracy results obtained over the period from January, 1990, to March, 1991.

In 1990–1991, North-west Airlines, Honeywell, and the Russian RIRT and AUSRIRE organisations carried out joint tests of airborne user equipment using both GLONASS and GPS. GLONASS showed latitude errors of 3.3–10.7 m and longitude errors of 2.3–3.4 m at a confidence level of 0.947.

There is a message transmitted by each satellite containing data concerning satellite health. This information is transmitted with a delay of not greater than 16 hours after detection of a fault. This is somewhat longer than GPS and is caused by the lack of truly world-wide monitoring stations.

2. GLONASS current status

During the first stage of GLONASS deployment, GLONASS satellites carried time/frequency standards with frequency stabilities of 5 parts in 10^{-12} per day and a life span of one year. In May 1985, the second stage of GLONASS deployment began. The satellites deployed during this stage were equipped with three caesium beam frequency standards having stabilities better than 5 in 10^{-13} per day. Satellite design life increased at first to two years and later up to three years.

During the period 1985–1987, the active life of the satellites was only five months to two years, but since 1989 life-span has usually been better than two years. Satellite operational reliability is being given considerable attention.

The GLONASS configuration, which has averaged ten to twelve satellites

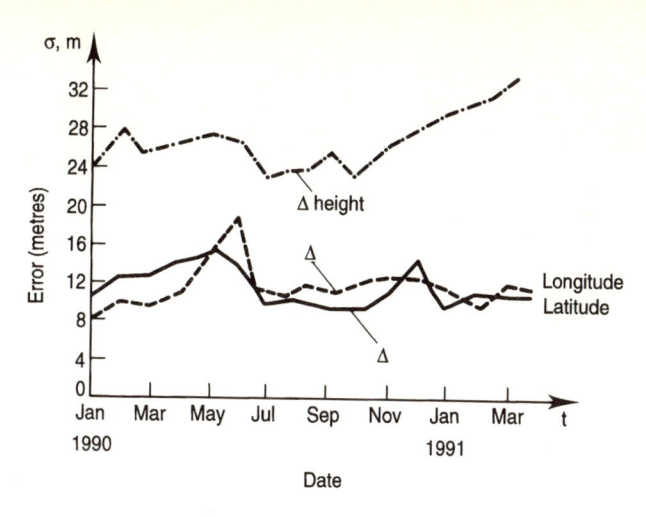

Fig 45 – GLONASS Accuracies

over the last few years, makes it possible for a European user to perform 3-d-navigation for 4–6 hours/day and 2-d for 20–22 hours.

At present, only one main measurement station is used in GLONASS control segment, using an ensemble of four hydrogen masers as the master frequency/time reference. The daily frequency instability of these hydrogen masers does not exceed 5 parts in 10^{-14}. Comparison of satellite time with this is performed using two-way laser retro-reflector measurements having an error no greater than 5ns (1σ). All GLONASS satellites carry laser retro-reflectors.

The availability of only one main measurement station for GLONASS frequency/time support hinders the development of the system, as well as degrading reliability of its operation. Unfortunately, putting a second station into operation has been delayed, since it was originally designed to be installed in Kazakhstan, now an independent state. Two command/measurement stations of the five stations used for orbital measuring are also now located outside Russia and their status is uncertain.

3. Prospects for GLONASS Development

Full-scale deployment of GLONASS comprising twenty-four satellites was originally scheduled for the end of 1995, based on the assumption that technical facilities would be available to launch two to three units per year, each unit comprising three satellites. However, considering the difficulties just mentioned, GLONASS full-scale deployment might not occur until 1996 or 1997.

Flight tests of a new GLONASS satellite called GLONASS-M are scheduled for 1996. The new features of this satellite are increased lifespan and use of a new spaceborne frequency/time reference with daily frequency instability no more than 1 in 10^{-13}. This will enable synchronization of satellite timing with an error not worse than 15ns (1σ) and will thus increase overall accuracy of the system. Along with this, GLONASS is planned to be used for the transfer of corrections for UTC (SU) − UT1-shift, as well as for that between GLONASS and GPS system times.

Chapter 7

Other Satellite Positioning Systems

While GPS and GLONASS offer continuous position-fixing for vehicles travelling at any speed and are world-wide there are a number of other satellite systems offering more limited fixing facilities. They may be limited in the area they cover; the type of vehicle that can use them; frequency of fixing; or in some other way, but provided their limitations are observed they can provide useful service. This is where the information given in Chapter 2 is valuable; given the number of satellites and their orbits it is often possible to say what their navigational usefulness is without having the manufacturer's data.

This is not an exhaustive list of all the satellite systems that have at one time or another promised navigational fixing capability but is a representative selection that have gone beyond the glossy brochure stage.

1. Inmarsat

INMARSAT is an organisation that was originally set up to manage the space segment of a satellite-based marine communications system which has been so successful that it is now a standard fitting on all large ships and many smaller ones. Inmarsat manages the space segment; that is, the satellites, but all ground uplink stations (Land Earth Stations, LESs) and the communications networks for them are managed by designated organisations in each country. In the UK it is British Telecom, in the USA COMSAT, and so on. Equipment on board ship is owned by the ship's management company but has to be licensed for use with the system.

Inmarsat has now spread to non-marine users. The bigger airlines are fitting their long-range fleets with it, for use by passengers as well as crew, and there are a growing number of land mobile and land-portable (news-gathering) users.

Inmarsat has always had in its charter the ability to provide the space infrastructure for satellite radio-determination, which includes both radio-navigation and radio-positioning, but required the authorisation of its member States to permit it to do so.

The civil aviation need to augment GPS/GLONASS with additional

The Air Pilot's Guide to Satellite Positioning Systems

integrity information broadcast from geostationary satellites has led to Inmarsat playing a major part in the provision of the space segment facilities for doing this, in the form of its specialised navigation payloads now due for launch aboard its Generation 3 satellites in 1996. As well as transmitting integrity data these satellites will also transmit a GPS-like ranging signal and eventually provide an elementary positioning service. It is not seen as a totally stand-alone service – the single-plane geostationary problem remains – but the net result, in a few years' time, might well be the nucleus of a civil-controlled satnav system using some of the features of the military systems but retaining a fall-back capability should those systems become unavailable. The idea has attracted considerable support in France, where it has become known as the 'European Complement to GPS' and has also gained the support of ESA (see later). In the following description the abbreviation 'GPSG' is used to indicate both GPS and GLONASS.

1.1 Method of Operation

The message transmitted from the navigation-payload satellites will, in the first place, carry integrity and differential data for GPSG derived from ground monitors. These monitors will be set up and operated by any country wishing to do so and will be linked into a world-wide GPSG integrity monitoring

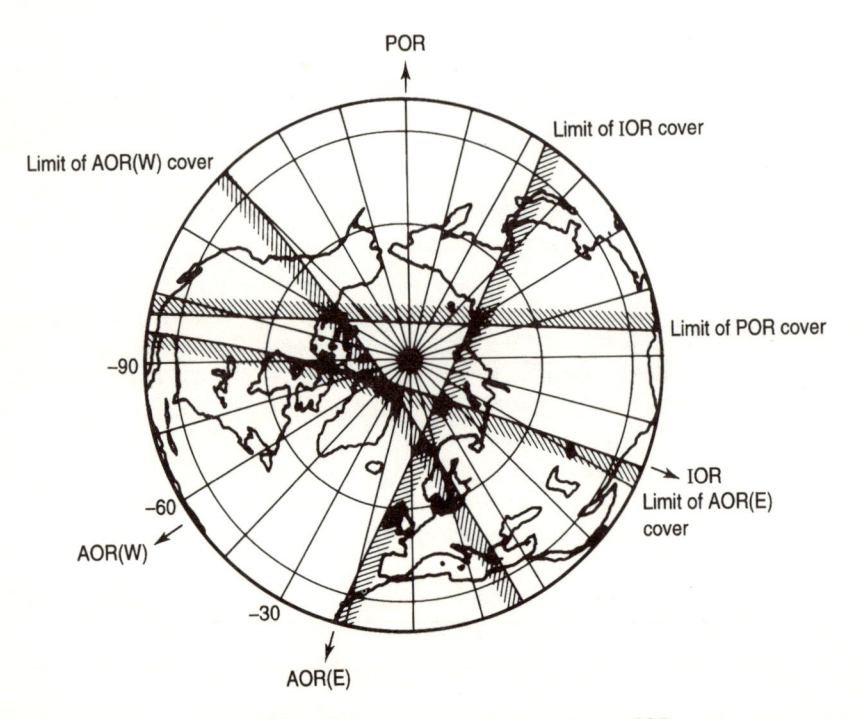

Fig 46 – Disposition of Inmarsat Satellites (1995)

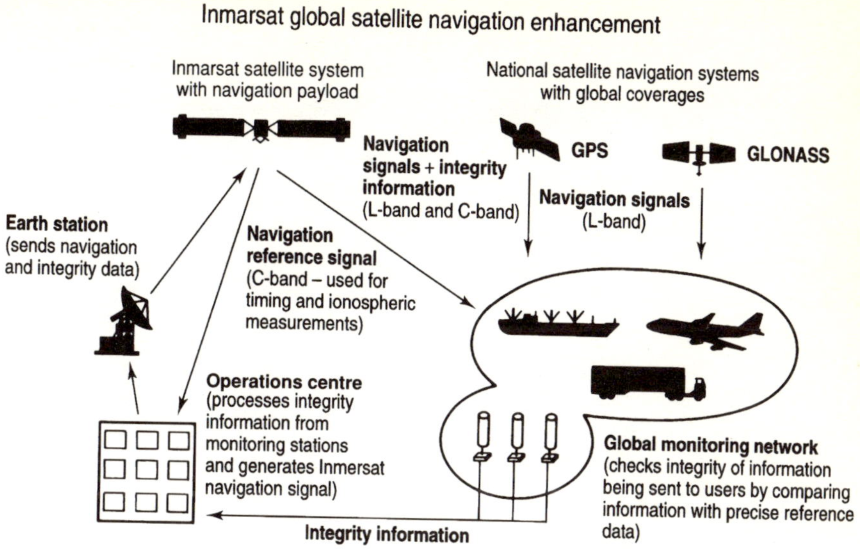

Other Satellite Positioning Systems

Inmarsat global satellite navigation enhancement

Fig 47 – Inmarsat GPSG Augmentation System

system. This system will be designed to provide independent checking on GPSG performance and will signal to aircraft immediately should there be anything wrong. The differential data will enable aircraft to increase the accuracy of their GPSG fixes, for instance, for precision approach. (For a fuller description, see Appendix 12, WAAS).

In addition, there will also be ranging capability so that GPS receivers can obtain extra ranges from them. This will be done by transmitting data and ranging in the form of GPS/GLONASS 'look-alike' signals. They will use radio frequencies either very close to GPSG or actually on the same frequencies and enough radio power to enable ordinary GPSG aerials to be used. GPSG receivers will see the Inmarsat transmissions as simply another set of GPSG signals and treat them accordingly, being differentiated from the 'real' signals by their code structure. The data message on them will be adjusted so that computation can be carried out with as little change as possible to the normal GPSG processing algorithms, and the user will get in effect a number of 'extra' GPSG satellites. This will offer considerable geometric reinforcement to GPSG, even if the user is mainly relying on GPSG.

Inmarsat navigation-payload equipped satellites may eventually become sufficiently numerous to be able to provide elementary fixing capability for non-critical navigation without assistance from GPSG.

This system would only require the addition of a few more satellites in non-geostationary orbits to complete a stand-alone system not requiring GPSG at all.

2. Geostar/Locstar

These systems were primarily mobile communications systems with positioning added, rather than navigation systems, and worked on similar principles. Although they were commercial failures, they illustrated a technique that has been re-proposed for other systems and are described for that reason. Geostar was the oldest, having been proposed in the USA in 1984 for use as an aircraft location system, while the technically similar Locstar system was promoted in Europe by a consortium led by CNES in France. They were co-operative systems requiring the user to transmit, but did not offer continuous fixing, nor was the fix necessarily available on board the vehicle, so they were not strictly navigation positioning systems and did not operate within the frequency bands allocated for navigation.

The following description of operation is limited to their positioning capabilities and uses Geostar as being typical.

2.1 Method of Operation

In the simplest form two satellites in geosynchronous orbit are used, either dedicated satellites, or other satellites with suitable payloads on board. A master control station transmits continuously a signal having suitable timing marks incorporated, through one satellite for reception by users. The user listens to this signal continuously but does nothing until he recognises his own digital identification being sent. He then transponds back a message relaying the timing information, the time delay caused by the turn-around having been carefully calibrated. This return signal is received aboard both satellites,

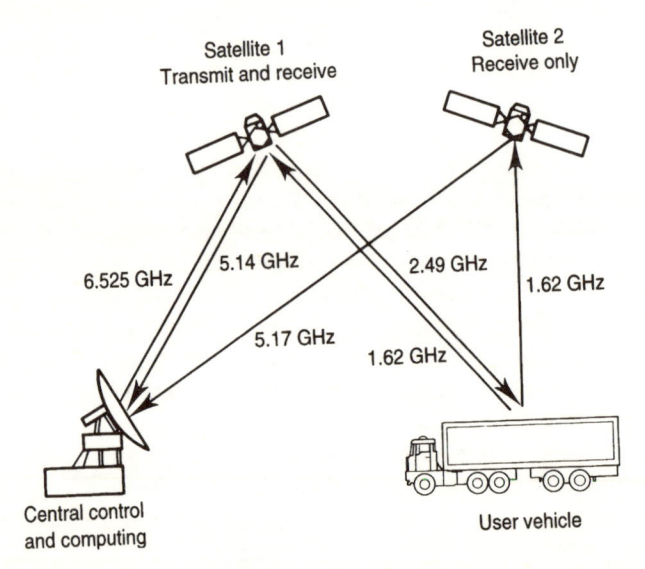

Fig 48 – Geostar System

which then relay it to the master station, which works out the user's position by the usual ranging algorithm for two-station fixing. In this way, there is no requirement for accurate absolute timing in the user equipment, which may be kept fairly simple. However, it has to be able to produce a transmission powerful enough to reach the two satellites and this is a restricting factor on size. If required, the user's position can be relayed back to him over the same link, but one of the major applications seen for this system was to enable fleet owners to keep track of where their vehicles were, and the positional information is needed mainly in their offices rather than at the vehicle. Other information might be relayed – type of load, new directions to the driver, and so on. The position so calculated does not need to be very accurate, ameliorating some of the problems that would become serious were high accuracy to be attempted. For instance, user's geocentric height is absolutely necessary as an input, as already described, and while the height of marine users might not be too difficult to derive, it is a much more difficult problem for land users, and even worse for airborne users. Even for an approximate solution, it is necessary to make some initial assumption about height, and then, having obtained the refined position, re-enter with a better height until the algorithm has converged to an acceptable accuracy. It was claimed that this could be done by using a digital height map of the areas covered, but since the height depends on the position and vice-versa there is some danger in certain situations of an unstable solution. An aircraft would have to transpond its own 'height' but unless it were fitted with a radio altimeter this would only be a pressure altitude and the height of the reference pressure surface above geocentric height would have to be calculated as well. It is in practice rather difficult to implement and there is scope for major errors to creep in. This, together with the rather low data rate, made it unlikely that it would be adopted for aircraft use, and probably contributed to its commercial failure.

3. Omnitracs/Euteltracs

Omnitracs (Euteltracs is the European equivalent) differs from Geostar in using K-band (transmit 14 GHz, receive 12 GHz) instead of L (1.5 GHz) and S-band (4 GHz), but is otherwise rather similar. There is a commercial advantage claimed for Omnitracs in that K-band frequency space is more easily available, particularly in Europe, than is anything lower. However, although a mobile antenna for K-band is smaller, more satellite power is needed to overcome rain and other attenuation.

The system uses transponders and aerials on standard communications satellites, providing a coverage pattern over Europe as shown in Fig 49. The reason for the restricted coverage is that the communications aerials on the satellites it uses are designed only to cover the populated areas of Europe and have a deliberately restricted radiation pattern. This has an advantage in that

The Air Pilot's Guide to Satellite Positioning Systems

Fig 49 – Euteltracs Coverage Pattern

because these are 'beam' aerials they produce a stronger signal than a wide-area aerial such as is used by GPS would do. In turn, a smaller aerial can be used on the receiving vehicle.

Euteltracs has five main elements:
* A customer terminal unit which enables users to send and receive messages and access position information about their fleet of mobiles. It is usually a standard PC running software provided by Euteltracs, which includes a map display for the immediate location of mobiles.
* A Service Providers' network management centre which processes data and keeps records of transactions with customers. It handles all the communications back and forth between the vehicle and customer. This is a packet network store-and-forward system and therefore there is a slightly indeterminate delay.
* A hub Earth station which does all the technical processing including receiving data from monitor stations and sending and receiving data from the satellites.
* Satellites. The standard Eutelsat satellites are used.
* A mobile communications terminal mounted on the vehicle which receives and transmits messages. The aerial is a small directional horn aerial which tracks the satellite being used on the basis of signal strength. The display unit incorporates a small keyboard for sending messages and can hold up to one hundred incoming messages. It has no voice capability.

Other Satellite Positioning Systems

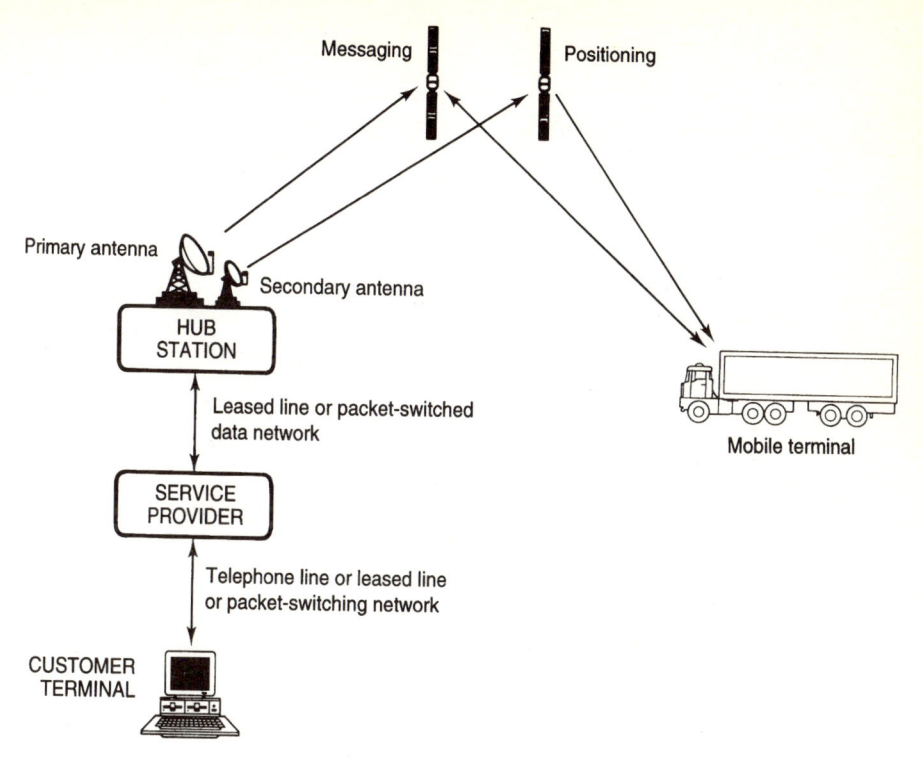

Fig 50 – Euteltracs System

Although its main purpose is communications for land vehicle fleet control purposes, it has a location capability which is performed in the same way as for Geostar. In the United States, initial positioning capability was provided by incorporating Loran-C receivers into the mobile units, but this is being superseded by GPS. In Europe, Euteltracs has achieved independent positioning capability by using both Eutelsats, but accuracy is not expected to be better than 500 metres.

For aviation use, there are a number of problems. The downlink power from the satellites is not sufficient for reception by an omnidirectional aerial and the Eutelsat mobile aerial probably could not cope with aircraft manoeuvring. Even if this problem were solved, the rate of position fixing is very slow for navigational purposes and probably could not handle aircraft speeds. In any case, Euteltracs does not claim to be a navigation service.

4. Iridium

The Iridium Corporation is owned by Motorola, Inmarsat and fourteen other telecommunications companies who have invested $1.6Bn in total. The system will use sixty-six satellites in low Earth orbit (750 km altitude) to provide world-wide cellular-network-like personal communications via small hand-held units. It is expected to be operational by 1998.

The satellites will be able to communicate with these hand-helds because they are at low altitude and will use beam aerials. Although the primary purpose of the system is communications, some positioning capability is claimed, although its accuracy is not stated in the current publicity material. From the navigational viewpoint, if the user can only communicate with one satellite at a time, as seems likely, only Doppler-shift could be used for location and since continuous transmission by the user is unlikely the accuracy would only be very low. If the user were in sight of three satellites simultaneously and triple ranging could be accomplished then fairly accurate positioning might be possible but there is no mention of this in Iridium literature. This type of system, although perhaps capable of intermittent positioning, would have little application in aircraft.

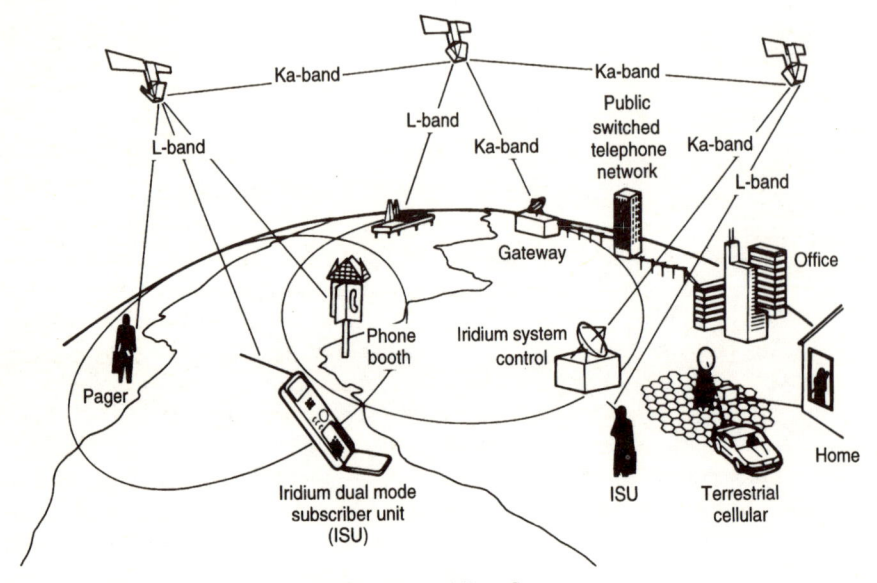

Fig 51 – The Iridium System

5. Orbcomm

Orbital Sciences Corporation owns this system, and in late 1994 bought the Magellan GPS receiver manufacturer, who will build the user handset for it.

Other Satellite Positioning Systems

Fig 52 – The Iridium Satellite System

It is a somewhat similar proposal to Iridium but using initially only twenty-six LEO (785 km high) satellites capable of providing 95% cover over the USA, which may be extended to other parts of the world eventually. Its primary purpose is data communication, with an aim of allowing users to send a 250-character message from anywhere in the world with a delay not greater than ten seconds. Unlike Iridium, positioning is specifically stated to be a feature of this system and will be done by a combination of Doppler measurements and GPS. The satellites will locate themselves by using on-board GPS receivers and relay their position to the user receivers, thus eliminating the need for expensive ground tracking networks. The user receiver, as in the old Transit system, will measure the satellite's Doppler-shift over a period of time and calculate its position in the same way. This, of course, will mean that it will suffer from the same problems as Transit (see Chapter 2) and is unlikely to achieve accuracies of much better than 500 metres. The system's advocates say that if a user needs better accuracy he can use GPS, and that a GPS set can be incorporated into their user receiver. The first satellite was launched in early 1995, with full capability scheduled for 1997.

6. Odyssey.

Like the previous systems, this is primarily a personal communications system with added positioning capability. It is owned by a consortium of companies comprising TRW, Teleglobe and ten other companies. It will consist of twelve satellites in medium Earth orbits at about 9,500 km (6 hours per revolution) but is not scheduled to be operational before 1999.

Chapter 8

Applications of Satnav in Aviation

As mentioned in Chapter 2, until satellites came along simultaneous high accuracy and long range were incompatible and many different radio navaids had to be developed depending on where the emphasis lay. The result has been a multiplicity of different aids all requiring their own specialist maintenance and support and the cost of running them is considerable. World-wide, about $400M/year is spent just on operation and maintenance without allowing anything for expansion and improvement. If satnav were adopted then many of these systems could be eliminated and much of this support cost saved. There is no doubt that even as it is, GPS can supply perfectly adequate accuracy for all en-route and oceanic flight and needs only the addition of integrity monitoring.

If satnav were installed for en-route flying then further benefits could be obtained if its capabilities were extended to precision approach and landing as well. The continued rapid expansion of air travel and the need for airlines to be able to operate reliably into smaller airfields has resulted in a demand for precision approach systems to be made available at more and more airfields. In the USA it is estimated that there are 1,046 runway ends that could benefit from the installation of precision approach and landing capability. Present plans are that the Microwave Landing System (MLS) will slowly replace ILS after 1998 and with its curved-approach capability there is no doubt it is technically suitable for the job. The difficulty is that MLS costs up to $1M per runway end to install while maintenance and calibration may cost a similar sum every year. In Europe, a similar programme of airfield improvement is under consideration, particularly in Eastern Europe and Russia, and the figures are not dissimilar. Studies are underway in both Europe and the USA to confirm the actual figures, but it is claimed that the adoption of satnav could halve maintenance costs, quite apart from the initial installation cost saving.

Compared with these figures, the cost of extending and improving GPS/GLONASS into a Global Navigation Satellite System (GNSS) becomes very realistic. However, these savings can only be made provided that satnav offers the same, or better, precision and reliability as the current ILS/VOR/DME systems. Both of these are critical to its success and are examined later.

There are other practical and non-technical advantages of satnav that are not always apparent. One is that satnav does not require, at least at its lower levels of accuracy, local installations or infrastructure on the ground. This could be of considerable value to third-world countries who wish to improve their navaids but have neither the funds nor expertise to implement complicated and expensive standard navaids. It would provide a rapid and low-cost method of upgrading their airfields and bringing more of them into use for airline operations.

Another is that it is sometimes very difficult to install conventional radio navaid transmitters in the remoter parts of the world. It is all too often found that the area in which they ought to be sited for best results is environmentally hostile with perhaps poor or non-existent access, no power supplies, and even physical danger for those who have to maintain them. Even in well-developed areas, satnav makes far less environmental impact than conventional systems. It does not require large antennas or installations and with ever-mounting complaints about the seemingly endless multiplication of radio masts and towers it is an argument that appeals to many.

These are convincing arguments for the adoption of satnav for almost all aviation purposes but they only apply in their entirety to a fully-developed civil satnav system such as GNSS will be. Until that comes about, the only available systems are GPS and GLONASS, neither of which were designed with civil aviation in mind. Both systems are deficient in important areas when civil use is contemplated and some of the work being done to remedy this is described in this chapter.

1. Integrity, Availability and Augmentation

Although interlinked to some extent, these are really separate questions. It is of course important to be sure that what the satellites transmit is always reliable, and systems designed to assist in this are designated 'integrity' systems. Such systems are not necessarily concerned with availability and could work perfectly well with only one or two satellites in view.

On the other hand, for universal useability the system must be available everywhere at all times. The original design objectives for GPS and GLONASS did not envisage having to provide 100% availability world-wide at the levels of accuracy required for civil aviation. Systems designed to remedy this are known as augmentation systems.

Note that integrity and augmentation systems are only needed because GPS and GLONASS were not designed to civil standards. If a new system designed from the outset for civil aviation were to come into being it would need neither integrity nor augmentation systems because they would be built into it from the outset.

1.1. Integrity

'Integrity' has been defined by ICAO as:

'The assurance that all functions of a system perform within operational performance limits'.

If a satellite transmitter breaks down there is not much of a problem; there is simply no information from that satellite and there is probably another one in view that can be used instead, although maybe with some slight lowering of accuracy. The dangerous situation is when a satellite is apparently working correctly but is transmitting faulty information. The ground control station could of course indicate this by changing the 'health' word, (always provided it knows there is a problem), but the way GPS is controlled at present may mean there is rather a long delay before that can happen. There is also the fact that while the satellite may not be operating well enough for some purposes it may be perfectly usable for others. Whether the GPS OCS considers a satellite unhealthy may depend on military mission requirements that may be irrelevant for civil users. This could result in a satellite being set unhealthy when there was no discernable problem for civil users, or, more dangerously, not being set unhealthy when it should have been.

The VOR, DME and ILS systems currently in use are closely monitored and have back-up equipment capable of being switched into use at a few moment's notice. They are on the ground, often at a major airfield, with engineering assistance available at short notice. That is not the case with the major component of a satnav system, the space vehicles. If anything major goes wrong the only solution may be to launch a new satellite, which cannot be done at short notice. To organise a new launch may take up to nine months, even when there is a spare space vehicle available. Partly, this is because navigation satellites have to be launched into specific orbital locations and the optimum times for launching into any particular slot may last only a few days and occur many months apart. Murphy's Law would ensure that the failed satellite was in the orbital position most difficult to access!

The design of the satellite constellation must take this into account, allowing a number of failures without affecting the performance of the whole system. This is, of course, why GPS has always been called a '21 plus 3' constellation – 21 satellites will do the job, the extra three being to allow the organisation of a new launch without loss of capability.

An obvious way of ensuring that the loss of one or two satellites would not have any significant effect on performance would be simply to launch a large number of satellites so that there would always be many spares always in orbit, but the cost and complexity of controlling them have to be considered. The number of 'spares' that would be required is highly dependent on satellite reliability – if they were totally reliable there would be no need for any at all! In fact, satellites, once in orbit, are indeed very reliable. The most dangerous part of a satellite's life is during its launch but if it gets into orbit working correctly then it is likely to go on working for a long time. Space is a relatively benign environment for electronics – there is no vibration and no moisture, and temperature is stabilised inside the satellite. The problems electronics in early satellites had with cosmic rays switching logic states have now been overcome. Transit satellites, the first navigation satellites, have lasted a long time – originally designed only for a life of five years several are still operating correctly after twenty years in orbit, and most of the Block 1 GPS satellites doubled their design life, the oldest now having worked for ten years. It may well be that the operational Block II GPS satellites will also exceed their design lives handsomely.

Be that as it may, considerations of forecast system-reliability on the best evidence available shows that GPS does not approach the standards required for civil aviation and an external integrity measuring system is needed.

1.2 Requirements for Integrity

There are two stages in an integrity system:

> 1. It must detect unfailingly whether anything in the system is outside specification. To do this means of course that it must itself have extremely high reliability.
> 2. It must warn the user with as little delay as possible. This may be as little as two seconds for some phases of flight.

To achieve (1) there must be a performance specification against which to measure it and in aviation this usually means the three-dimensional fix accuracy. The required fix accuracy varies depending on what the aircraft is doing, (Table 11), so the system must allow for this. Unacceptable accuracy for precision approach might be quite satisfactory for en-route flying so the system cannot simply be declared totally unusable the second anything goes out of specification.

In this table, 'source accuracy' is the accuracy of the basic source of navigational information. 'System Use Accuracy' is the accuracy with which it is expected aircraft will be flown in practice when using a source of the specified accuracy, taking into account interpretation and pilotage errors.

GPS provides a source accuracy of 100 metres at the 2drms level, and is therefore technically acceptable for all phases of flight down to non-precision approach, beyond which it needs augmentation (see later).

1.3 Ground-based Integrity Systems (GIS)

Integrity determination based on fix accuracy as measured at a ground station is complicated by the fact that only four satellites are needed for a fix although as many as ten or eleven might be visible. Different makes of GPS receiver may select different sets of four, or maybe use six, or eight, or all of them and there is no way a monitor can know this. Even if a ground monitor calculated all the fixes obtainable from all possible combinations of satellites it still could not tell an aircraft not to use a particular combination because the geom-

Table 11: Controlled Airspace Navigation Accuracy to Meet Current Civil Requirements

Phase	Sub-Phase	Altitude FL/FT (AGL)	Traffic Density	Route Width (nm)	Source Accuracy 2drms (Metres)	System Use Accuracy 2drms (Metres)
	Oceanic	FL 275 to 400	Normal	60	N/A	12.6nm*
		FL 180 to 600	Low	16	2,000	7,200
	Domestic		Normal	8	1,000	3,600
		500 FT to FL 180	High	8	1,000	3,600
En Route/ Terminal	Terminal	500 FT to FL 180	High	4	500	1,800
	Remote	500FT to FL 600	Low	8 to 20	1,000 to 4,000	3,600 to 14,400
	Special helicopter operations	500 to 5,000 ft	Low (off-shore)	Not determined	1,000 to 2,000	3,600 to 7,200
		500 to 3,000 ft	High (land)	4	500	1,800
	Nonprecision	250 to 3,000 ft	Normal	N/A	100	150
	CAT I	N/A	Normal	N/A	±17.1** ±4.1***	N/A
Approach and Landing	Precision CAT II	N/A	Normal	N/A	CAT 1 Decision Height Point **** ±5.2** ±1.7***	N/A
	CAT III	N/A	Normal	N/A	CAT II Decision Height Point **** ±4.1** ±0.6*** At Runway Threshold ****	N/A

* The distribution of this error is detailed in the 'Report of the Limited North Atlantic Regional Air Navigation Meeting,' dated 1976; ICAO Montreal, Canada.
** Lateral position ground equipment (2 sigma) accuracy in metres for Precision Approach and Landing.
*** Vertical position ground equipment (2 sigma) accuracy in metres for Precision Approach and Landing.
**** Assumes a 3° glideslope and 8,000 ft distance between runway threshold and localizer antenna.

etry (DOP) being experienced at the aircraft might be different from that at the monitor if there were several hundreds of miles between them. Integrity monitoring by fix accuracy is therefore inadequate.

The solution is to monitor individual satellites by comparing actual pseudo-ranges with those expected. The data message is also examined to make sure that no false data is being transmitted. The ground station decides whether anything is outside specification and over a data link tells aircraft what it finds, leaving it up to them to decide what to do. It is not often that pseudo-ranges suddenly jump, the most common form of problem being a slow timing drift which produces a slowly increasing error in the pseudo-range measurements. Although it is complicated by the deliberate Selective Availability time-jittering of GPS, the random nature of S/A makes it different from this type of slow continuous drift.

1.4 Receiver Autonomous Integrity Monitoring, RAIM

Ground Integrity Systems need a continuous data-link to the aircraft, introducing integrity problems of its own, and therefore efforts are being made to develop integrity checking within the aircraft receiver itself that does not need outside assistance. These systems rely on the fact that although only four satellites are required for a fix, more are usually available and therefore they can be used to form fixes from different combinations of four that can be intercompared for consistency. Theoretically, five satellites enable a receiver to determine whether anything is wrong, but not where the problem is, while six satellites will determine which satellite is the problem. However, it depends on all the satellites used being in good geometrical positions (good PDOP) and it is often the case that some of them are not sufficiently well oriented to be useful. So it might be that even although eight or nine satellites are being tracked, RAIM is still not possible because of geometry.

RAIM is really only a specialised 3-D example of the elementary navigational principle explained in Part 3 of Chapter 1 where (in 2-D systems) a fourth position line is needed to resolve which of three is wrong. Satnav is 3-D and the triangle described there turns into a tetrahedron when there are four satellite ranges. If there are three satellites fairly low down spaced at 120° intervals around the horizon and one more vertically overhead it would be a pyramid, and if all the ranges were wrong by the same amount then the pyramid would get bigger but the centre would stay where it was. If one range only was wrong then it would get bigger and the centre would move but there would be no indication as to which one was causing it, so RAIM cannot work with only four satellites.

With five satellites, there are five possible combinations of four-satellite fixes and provided all combinations have the same PDOP one will produce a smaller tetrahedral volume than any of the others. This is the one that does not contain the faulty satellite, but all the others do, so it can then be isolated. Unfortunately this depends not only on the same PDOP requirement, which rarely occurs, but also on what is assumed to be an unacceptable tetrahedral volume – obviously no set of satellites is going to make it zero and just because one is a bit smaller than the others does not necessarily mean there is a faulty satellite. This makes fault detection with only five satellites problematical – some have said impossible. If there are six, the chances rise significantly, and with seven even better. However, even if there were eight, (the average maximum in view at any one time above 10° with a twenty-four-satellite constellation) it has been calculated that the chances of correct detection and isolation are still only about 65%, as against the 99.9% certainty demanded for the critical approach phases of flight if GPS is to be used as a sole-means navigation system.

A modification of this method is to relax the requirement to positively identify which satellite has failed. If it is sufficient to be sure it is one of two possibilities, and both are excluded, then it has been claimed that the detection rate rises to 86%, but this is still well below what is needed and in any case positively requires there to be eight satellites in view.

This problem leads to a consideration of how many satellites are needed to ensure that RAIM always works anywhere at any time and not surprisingly the figure is quite high, coming out at between 36 and 42, considerably more than GPS's 24. If GPS and GLONASS could both be combined into a common system then there would be forty-eight satellites available which would certainly solve the problem, and this is one reason why combined GPS/GLONASS receivers are being developed. Also, the transmission of ranging signals from satellites used for integrity data dissemination is useful in this respect and is why WAAS (Appendix 12) will be using them.

1.5 Availability and Augmentations

An integrity system's function is only to say whether the satellites are working properly, and of itself does nothing to get over the second problem inherent in GPS/GLONASS not being designed with civil aviation in mind. This is that only the minimum number of satellites were launched in those systems to satisfy military operational requirements. It was decided that basic military missions could be carried out if there were only twenty-one satellites operational at any one time and that an adequate constellation would therefore be twenty-four satellites, allowing three to fail. Using the design reliability of the satellites, this gave a 98% probability of twenty-one satellites operational at any one time. This is far below civil requirements, which are that a system must be available for 99.998% of the time at the specified accuracy level. Not only this, but because satnav is necessarily world-wide, this figure must be maintained everywhere in the world. The number of satellites visible at any one time affects only one thing – geometry, a main determinant of system accuracy, which is expressed mainly by the PDOP number. If it is determined that a minimum PDOP of, say, four is required for civil operations and that this must be met twenty-four hrs/day everywhere in the world it turns out that the number of satellites needed is much greater than twenty-four, in fact, something like forty. It is not an accident that this is very like the number needed for RAIM to work properly – the reason it does not is mainly due to inadequate geometry. The extra satellites do not have to be in the same intermediate Earth orbits as GPS; the only requirement is that they are on the same time-base . If GPS and GLONASS could be combined they would do the trick if put onto the same time-base but at present they are not. Another way of tackling the problem is to use geostationary satellites, and this was another reason for placing navigation payloads on the Generation 3 Inmarsat satellites. A full description of the proposed Inmarsat-based GPS augmentation was given in the previous chapter.

2. Airfield Approach and Landing

Accuracy and integrity are of even more importance than during en-route flying where the safety of a landing aircraft is concerned. The standards set for the existing systems, ILS and MLS may be used as guides to what is needed.

2.1 ILS

ILS is the primary world-wide, ICAO-approved, precision landing system. It was originally designed as far back as 1944 and although well proven is now conceptually out-of-date. Scanning beam MLS has been developed to replace it, but is itself a twenty-year old system and thought by some to be outdated even before it is generally installed. However, it has a considerable advantage over ILS in that it is a wide-area system covering a wide angle either side of the runway and not just providing a single high-accuracy beam as does ILS.

ILS course alignment (localizer) accuracy at runway threshold is maintained within ± 25 feet. Glide-slope course alignment is maintained within ± 7.0 feet at 100 feet (2 sigma) elevation.

A very important feature of ILS is its built-in monitoring system. If anything goes out of tolerance either or both localiser and glide-slope can be shut down very rapidly. The specified shutdown delays are:

	Localizer	Glide-slope
CAT I	</= 10 sec	</= 6 sec
CAT II	</= 5 sec	</= 2 sec
CAT III	</= 2 sec	</= 2 sec

2.2 MLS

MLS azimuth accuracy is ± 13.0 feet (+ 4.0m) at the runway threshold and elevation accuracy is ± 2.0 feet (+ 0.6m). Ranging accuracy is ± 100 feet for the precision mode and ± 1,600 feet for the non-precision mode.

Elevation angle is transmitted at 39 samples per second, azimuth angle

Applications of Satnav in Aviation

at 13 samples per second, and back azimuth angle at 6.5 samples per second. The airborne receiver averages several data samples to provide fixes of 3 to 6 samples per second. A high rate azimuth angle function of 39 samples per second is available and is used when there is no need for flare elevation data.

MLS integrity, like ILS, is provided by an integral monitor. The monitor shuts down the MLS within one second of an out-of-tolerance condition. These are the standards a GNSS must meet or exceed if it is to be considered as a possible landing system.

2.3 GNSS as a Landing Aid

Since GNSS does not exist as yet it is difficult to assess how it might perform in the landing role, but GPS has been extensively evaluated and can be used as a guide. The one major difference between them is that presumably GNSS would not have SA and would therefore start out with a much higher integral accuracy than GPS.

2.3.1 General Advantages

GPS is not a beam system and having no azimuthal coverage limitations it can be used for any azimuth of approach without any special arrangements. Also, its accuracy is unaffected by distance or height. This is of great importance for helicopter approaches in that there is no need for rigid approach patterns to be established unless there are reasons having nothing to do with navigational accuracy. It could be used even if the landing platform was moving, e.g. landing on a ship.

GPS radio-signal availability would be very high. The radio frequencies it uses are comparable in performance with those of MLS, and well above those of ILS. It suffers no problems from rain static, night effects, thunderstorms, and so on.

Its accuracy is maintained twenty-four hrs/day. It is virtually the same at all times and does not vary between night and day. The only variation in satnav accuracy is caused by geometric effects.

2.3.2 GPS Accuracy Enhancement for Landing

If we take MLS as a guide, GPS must produce accuracies of 4m horizontally and 0.6m vertically. Even to meet current ILS standards it must produce 7m and 2m respectively. In its standard navigational mode it simply cannot do this, even were the full undegraded PPS to be made available to civil users. A full GNSS designed for maximum accuracy from the outset might just do it unaided, although even then the vertical accuracy is questionable. Clearly, some other method of increasing accuracy is needed.

An obvious method is to use differential techniques (described in Ch. 5). Marine DGPS trials have shown that it is not at all difficult to provide an improvement in horizontal accuracy to 5m or even 3m using only standard techniques. The newer techniques such as wide-area combined corrections can improve on this to the 2–3 metre level.

Even this is not quite good enough and this has led to a search for further improvement. Marine differential systems work only in the 'code-only' mode and a considerable improvement can be made by measuring in addition the phase of the carrier wave, as described in Chapter 2. The extra phase-derived data must be sent to the aircraft over the same data link as is used for code corrections, and this makes considerable additional demands on it. However, the real problem is that there is still no way of telling which cycle is correct. It is exactly the same as the old problem of lane ambiguity with Decca Navigator/Loran-C.

One method of resolving this problem makes use of the very high speed of computation now possible even in small computers by simply computing every possible position from every possible set of satellites within the radius of the accuracy of the code-derived fix, accepting the one with the smallest dispersion as correct. What this means in practice if there are six satellites visible is that there are some 3,000,000 position solutions to be calculated every second!

Another way is to make the accuracy of ordinary code-tracking sufficiently high to be able to resolve the 19 cm ambiguities itself without outside assistance. This was described in Chapter 5.

Yet another method of improving accuracy is to install a pseudo-lite at the landing site. A pseudo-lite is a differential monitor station that transmits its data over a spread-spectrum data link at a frequency near to that of GPS. Its signal is deliberately made as like GPS as possible and its frequency is close enough to GPS for a standard GPS receiver to pick it up, but not so close as to jam it. In this way, an additional range is obtained directly from the landing site that can be processed in parallel with the real GPS signals in a GPS receiver.

A development of this idea is the Integrity Beacon Landing System (IBLS) developed by Stanford University in the USA (Fig 53).

In this system, two very low-powered pseudo-lites are sited one each side of the approach path about half a mile from the touchdown point. They are locked in code and phase to the GPS system time-base. The combined effect of these two additional 'satellites' on the real system is that any cycle ambiguities in the aircraft receiver are resolved as it flies through them, thus correcting the aircraft system to an accuracy of a few centimetres in time for the final touch-down. Test flights have shown that this system is capable of supporting accuracies of 30–50 cms in three dimensions (Fig 54).

The potential of these systems for landing is shown in Table 12, a summary of the results of flight tests carried out by the FAA using a number of different techniques and receivers (except IBLS). The ability to achieve one metre accuracy in both horizontal and vertical planes is clear.

However, these results were achieved by using highly specialised equipment under test conditions and there is still some way to go before they can

Applications of Satnav in Aviation

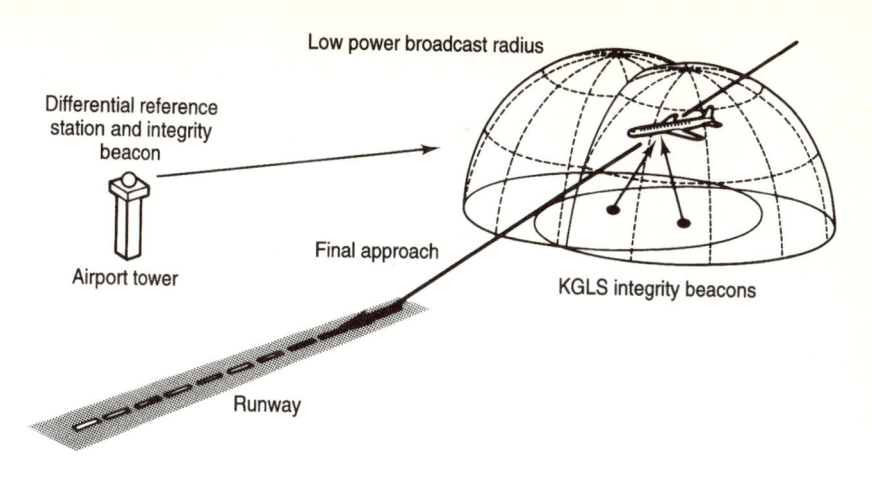

Fig 53 – The IBLS

be realised in practice as an everyday event. Also, proving integrities of perhaps one in ten million requires a vast amount of testing and it will be some time before it is completed.

3. Integrity and Augmentation Systems Combined

An integrity system has to measure pseudo-range errors in order to determine integrity and it will not have been lost on the reader that this is exactly what differential stations do. It is only logical to transmit these errors together with the integrity information and make a combined system that besides guaranteeing integrity also improves system accuracy. Because of this improvement in accuracy there is also some degree of availability improvement, but of course the basic problem of lack of satellites is not cured.

The data link for any of these aviation systems has to be a satellite data link in order to get both reliability of the link itself and wide-area oceanic coverage. Usually, a geostationary satellite is used but this is not essential – if there are enough of them, satellites in other orbits could be used. One of the most efficient methods of transmitting data over a satellite is to use a spread-spectrum modulation system, and in turn this means that ranging can be performed on it. It only needs the signal timing to be locked to that of GPS to supply an additional ranging measurement that can be combined with GPS to improve GDOP. This is the basis of the Inmarsat GPS augmentation and has been taken a further step in the US WAAS proposal (summarised in Appendix 12), where the use of other satellites in addition to Inmarsat is foreseen.

When WAAS is in operation, by 1997, there will be another four or five satellites providing ranging signals as well as integrity and differential information.

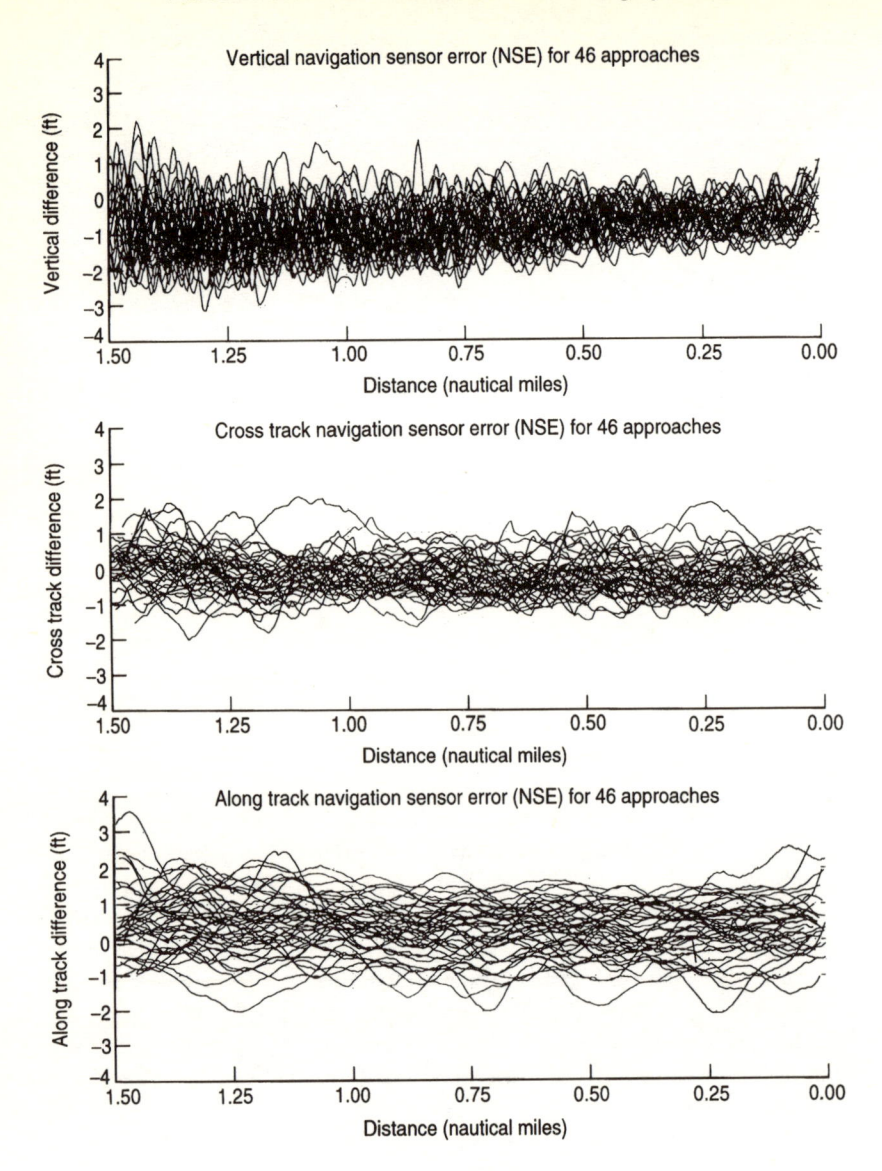

The Air Pilot's Guide to Satellite Positioning Systems

Fig 54 – IBLS In-flight Results

It is only one further step beyond this to add navigation payloads to other satellites, or even launch satellites specially for augmentation purposes and the extra satellites needed to put GPS onto a sound basis for civil users will be there. This is some time in the future but it is encouraging that there are plans already being developed to do this.

Table 12: Summary of Flight Tests

Receiver	Augmentation	SA	Differential Up-date Rate	Position Up-date Rate	|M|+2SD Vertical Sensor Error, Final 2 nmi	|M|+2SD Crosstrack Sensor Error, Final 2 nmi	Average VDOP	Number of Approaches
2-Channel Sequential Receiver (SEQ)	none	Off	N/A	1 Hz	12.7 m	7.7 m	1.7	18
"	IRS	Off	N/A	10 Hz	8.5 m	8.1 m	1.7	8
"	Differential	On	1/2 Hz	1 Hz	11.2 m	5.9 m	2.0	33
"	Differ/IRS	On	1/2 Hz	10 Hz	8.4 m	4.3 m	2.0	33
6-Channel Carrier Aiding (CAID)	Differential	Off	1 Hz	2 Hz	5.1 m	4.7 m		18
"	Differential	On	1 Hz	2 Hz	7.0 m	6.1 m		10
10-channel Narrow Correlator Spacing (NCOR)	Differential	On	1 Hz	5 Hz	2.1 m	1.2 m	1.7	35
On-the-Fly Kinematic (KIN)	Differential Phase Measurement	On	1/2 Hz	2 Hz	1.0 m	1.0 m	1.9	18

Appendix 1
Methods of Obtaining Range by Radio

1. Pulse Systems

The simplest radio signal is an unmodulated continuous carrier wave which unfortunately provides no information apart from the fact that it is there. About the only thing it can be used for is to provide a bearing, but as we have seen, this is impractical if the transmitter is on a satellite. To carry out ranging the radio signal must have identifiable timing marks and a plain carrier wave, simply because it is absolutely continuous, does not. If instead of being continuous it is switched on and off very rapidly in short bursts (pulses), the necessary timing marks are provided by the start and stop of each pulse. If they last only a few millionths of a second each, and enough time is allowed to let each pulse get to a target and be reflected back again before the next one goes out, then the time lapse between transmission and reception represents twice the range to the target. This of course is radar, which has the advantage that the target need be nothing more than a reflecting surface. A corner reflector is a good 'beacon' for this purpose and can often be seen on airfields for calibrating radars, or carried on small boats to enhance their radar signature. Sometimes the reflected wave is too weak, or more range is required, and an active transponder that re-transmits a much more powerful signal back to the receiver is used, as in secondary surveillance radar, SSR.

1.1 Transponder-assisted Ranging Methods

When the transmitted signal is passively 'bounced' back to the receiver, there is a great loss of energy at the reflection point and reflection efficiency might be very low. To ensure there is always a usable strength of reflected signal back at the receiver there has to be a considerable amount of power in the radiated pulse. The problem is far worse if the target is a comparatively small satellite 12,000 miles away and usually means that satellite-tracking radars have to use high-gain aerials and high transmitter power, so it is really not feasible aboard ordinary aircraft to use simple radar-ranging to obtain range to a satellite. However, if all that has to be done is to get an acceptable signal up to a satellite one-way in order to trigger-off a satellite transponder the problem becomes more manageable.

The other problem is the stability of the timing system in the receiver. Round-trip ranging systems, whether transponder-assisted or not, rely on the receiver time-base remaining accurate for the duration of the round-trip. But a quarter of a second (the time needed for a radio wave to get to a geostationary satellite and back) is not very long and if only moderate accuracy is needed then fairly ordinary electronic engi-

Appendix 1

neering techniques offer sufficient stability although for really high accuracies more specialised techniques may have to be used. To obtain the right geometry for a fix, ranging must be conducted to several satellites simultaneously, but this is not difficult to arrange provided enough satellites are there. Descriptions of systems using these techniques are given in Chapter 7.

The generic name for systems ranging simultaneously to two transmitters is rho/rho, from the Greek letter ρ. Some systems permit simultaneous ranging to three transmitters and not surprisingly are called rho/rho/rho systems.

1.2 One-way Range Measurement

It would be a lot better if there were no need to rely on reflections or transponders. The ideal would be a satellite transmitting a ranging signal that only needed the user to have a receiver but the basic need for timing marks would still be there. In order to measure range, the receiver must have a starting point to trigger-off its timing at the exact moment the satellite transmits, but since nothing can travel faster than a radio wave the satellite could never tell the receiver itself. However, if range between user and satellite could be reduced to zero, or happened to be known some other way, perhaps by knowing exactly where both satellite and user were by using another navaid, then the known range could be turned into time and the difference between the satellite and receiver clocks determined. This may sound a bit silly – after all, the purpose is to measure range and if range is already known why bother? – but it would have its uses. For instance, an aircraft could fly (very low!) over the beacon so as to make the range as close to zero as possible and synchronise its receiver clock as it did so. Then, as it flew away, its receiver would (hopefully!) stay in synchronism and could provide its own start point for timing. Continuous ranging from the beacon would be available thereafter. The problems are obvious – it would be physically impossible to visit all the different transmitters that might be used (their clocks would inevitably follow Murphy's Law and be working on different times) and the receiver clock would have to be extremely stable. A good crystal oscillator might be out by 10uS at the end of a day, and in 10uS a radio wave travels three kilometres. Atomic clocks could improve on that quite considerably, and they are in fact used in some types of offshore survey systems, but their expense is prohibitive for normal users.

Therefore, without either a transmission from the 'receiver' (as in *1.1*); or calibration at a known point, direct ranging cannot be performed simply because the timebase in the receiver cannot be locked to that of the transmitter. The consequence is that no satnav system designed for continuous fixing has been proposed using only a single satellite.

2. Range Measurement Using Carrier Phase

In Fig A1 a transmitter raises the electrical power fed to an aerial from zero to a maximum, back through zero to an equal but negative amount, and finally returns to zero.

This is one cycle of energy and the time taken for it is determined by the radio frequency – if the radio frequency is 1 MHz (one million cycles per second) then each cycle takes one-millionth of a second. Radio energy flows out from the aerial at the

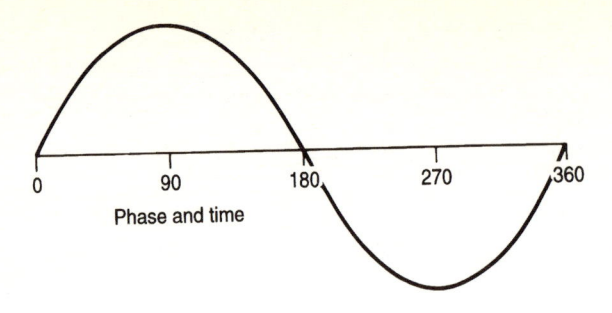

The Air Pilot's Guide to Satellite Positioning Systems

Fig A1 – RF cycle

speed of light, and is replaced by further energy from the transmitter during the next cycle. The shape of this cycle of energy is a sine-wave and the point along it reached by the RF energy is called its 'phase'. Like pulse-timing, phase must be measured from some reference point such as zero phase. When the cycle is complete and phase is once again zero, the first part of the cycle will be a considerable distance away. If the frequency is one million cycles per second (1 MHz) then, because the speed of light is 300,000 kms per second, the RF energy emitted at the start of the cycle will have travelled one-millionth of 300,000 kms, or 300 metres, by the time the cycle is complete. The wavelength of a frequency of 1 MHz is therefore 300 metres, illustrating the interchangeability between frequency (time) and wavelength (distance).

If a CW transmission could somehow be turned around by the user's receiver and sent back to its origin, then a phase comparison could be made at the transmitter between the outgoing and incoming signal. Provided the user were not more than half-a-wavelength away it would then be possible to calculate his range and it would not be very different from using pulses, but there are severe practical problems. How would the receiver, with a transmitter next to it which would still be transmitting, be able to receive the weak return signal on the same frequency? And to get any reasonable range very low frequencies with long wavelengths would be required. For instance, if the maximum range was to be only 100kms then one wavelength would have to be no shorter than 200kms otherwise it would not be possible to tell which cycle any particular phase was referred to. 200kms represents a frequency of only 1.5 kHz, which is extremely difficult to transmit as a radio frequency. However, there was one system that worked along these lines, designed in Germany in 1939. Instead of trying to send out this low frequency directly as a radio wave, it was used to modulate a high-frequency carrier wave that could be radiated easily. An aircraft picked it up and transmitted the low frequency back using a different carrier frequency. The phase of the low-frequency modulation was preserved during re-transmission and the distance measurements were carried out by comparing the phase of the returned modulation with that going out. The outgoing and incoming signals could be separated because they were being carried on different radio frequencies. The idea was never developed because simultaneous use by many users is difficult; every user must have a transmitter and thus a different radio-frequency; and maximum range is low. Although it has had some use in special-purpose telemetry systems it has never since been employed in navigation systems.

A system using crude phase in this way is impractical but techniques in which the phase of one transmitter is compared with that from another are quite practical and have been used in many systems such as Decca and Loran. These are described later.

3. Differential Range Measurement

The problem of synchronising a receiver timebase with that of a transmitter without a link between them is insoluble, but if several transmitters are all on a common timebase then a receiver can measure *difference* of range between them without any need to synchronise itself.

Transmitter synchronisation is done by arranging for one transmitter to pick up the signal directly from another and use it for timing its own transmission. Any drift in the first signal will be followed by an exactly similar drift in the second, keeping the *difference* between them always the same. Several transmitters might synchronise themselves to one central transmitter in this way and this is where the 'Master' and 'Slave' terminology used in some systems originated. The principle works because the need for 'absolute' timing has been eliminated and all that is being measured is difference. It is a powerful technique and can be used at any radio frequency. It was first applied at low frequencies (100 kHz or so) for systems that were required to provide good over-sea ranges but has become a standard technique in all high accuracy navaids.

4. Position-fixing Systems

As already seen, a single position line, whether bearing or range, does not provide a fix; at least two are needed. Straightforward ranging systems like DME, that require the user to transmit, can provide a fix from only two beacons, provided geometry is correct, but differencing systems need three transmitters; in effect, one transmitter provides a timing reference for the other two. If you like, the problem of providing an absolute timing reference has been taken away from the user and given to another transmitter.

When transmitters are locked together in this way direct measurement of range is still impossible because the receiver itself is not part of the timing loop. Its own internal timing will have an unknown offset that cannot be taken out. However, *difference of range* between two transmitters can be measured easily enough by simply measuring the time delay between reception of them using one as a start point. If a line is then drawn on a chart joining all points having the same time difference, it will be a hyperbola, hence the name 'hyperbolic' given to systems of this type (Fig A2). Two transmitters are needed to provide one position-line but since one transmitter can act as a reference for several others only three are needed for two position lines.

Many systems have been designed on these principles. Often, one station is designated as a 'Master' station, and several other transmitters are arranged around it to act as 'slaves'. Differences are taken between the Master and each slave, resulting in multiple position lines which provide the fix.

If the transmitters radiate pulsed signals then the pulse repetition rate is made sufficiently slow that no user, even at maximum range, receives two pulses from a distant

A = Master
B, C, D = Slaves

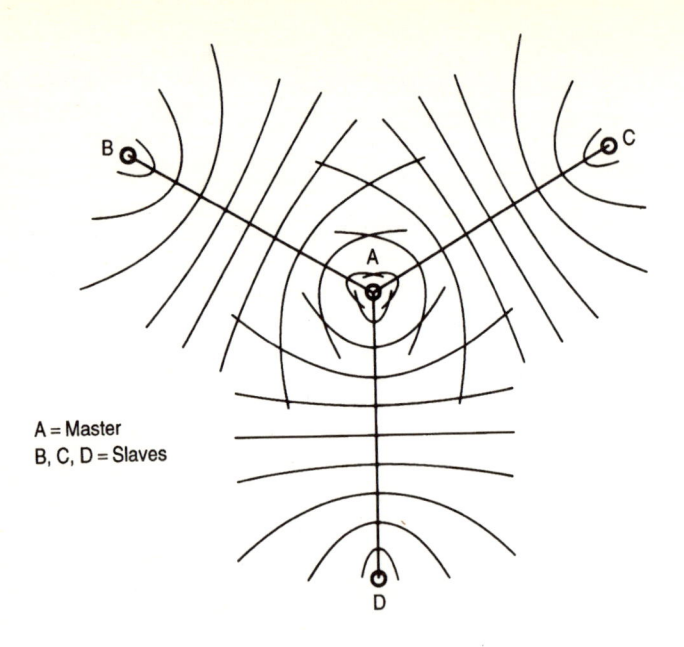

Fig A2 – A Hyperbolic System

transmitter while getting only one from a nearby station, because this would result in ambiguities. But if CW transmissions are used and the timing measurement is done by phase comparison then even at low frequencies there will be multiple ambiguities that have to be resolved in some way. The latest versions of GPS receivers used for surveying have returned to the phase-comparison idea to increase their accuracy, and face the same problems of ambiguity.

5. The Ambiguity Problem

In CW systems, because every cycle is like every other cycle, the same phase readings occur every wavelength, and when they are differenced similar phase differences occur every half-wavelength. Only if the coverage area is so small that it is completely within the distance of half a wavelength is the problem irrelevant, so some way of telling one from another must be incorporated. One way, used in Loran-C, is to break up the CW transmission into a series of rather long pulses and use, in effect, the coarse pulse time-difference to resolve which cycle is correct. Each Loran-C 'pulse' contains about twenty-five cycles and the accuracy of a straightforward pulse envelope measurement would be quite insufficient so actually only the leading-edge of the pulse is used. The difficulty this causes is that what was previously a narrow-band system becomes broad-band, occupies more spectrum, and needs high-power transmitters. The other way, used in Decca and Omega, is to use a combination of approximate knowledge of position together with a way of periodically producing

Appendix 1

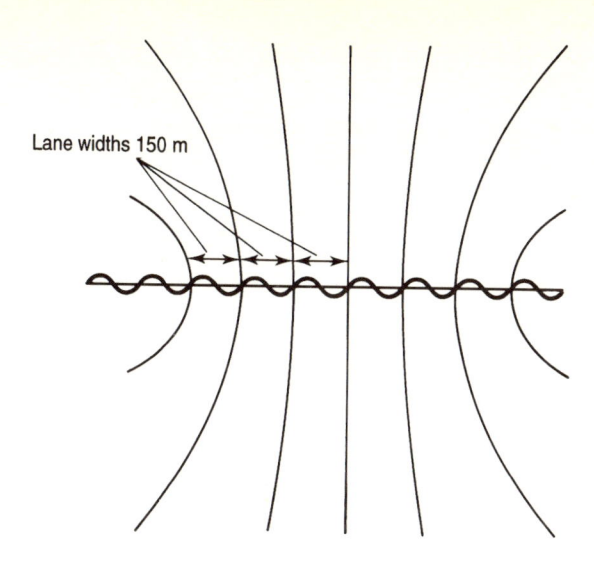

Fig A3 – Hyperbolic Ambiguities

lower frequencies that are capable of resolving the ambiguities of the higher frequencies.

In Fig A3 a practical hyperbolic system using a fundamental frequency of 2 MHz is shown producing multiple lanes (and ambiguities) 150 metres apart. A typical phase-measuring accuracy of 10° would produce a ranging accuracy of 1/36th of 150 metres, or about four metres. This is excellent basic accuracy but the vessel using the system would have to know where it was within 150 metres to start with. To get over this, at regular intervals a lower frequency of 0.2 MHz (200 kHz) might be produced from the same transmitters giving a lane ten times longer, 1,500 metres. Each new lane would thus contain ten of the narrow ones and if measuring accuracy were, as before, 10°, then positional accuracy would be 1/36th of 1,500 metres or forty metres. This is quite sufficient to resolve which of the narrow lanes is correct, although probably not good enough for the work in hand. Now, the vessel need only know where it is initially to within 1.5 kms instead of 150 metres, a much more manageable figure.

It might seem that by extending this idea to lower and lower frequencies all ambiguities could be eliminated but it cannot be carried too far because of engineering and propagation problems. It is just about feasible to design a transmitter and aerial system capable of switching every few seconds between 2 MHz and 200 kHz, or transmitting both simultaneously, but any bigger difference would be impossible. Also, the difference in propagation characteristics between frequencies more than 10:1 apart in frequency is sufficiently great to introduce differential phase delays that destroy the necessary phase coherence.

There are many systems using this basic idea in different ways. Decca Navigator is probably the best known, arranging for its transmitters to periodically swap frequencies in a pattern that allows derivation of a family of frequencies to be used for lane identification.

All of these systems are designed primarily for marine users and the reason why they are not much used by aviators is fairly obvious. Obtaining an initial position to within 1,500 metres is not difficult for a slow-moving ship, but for an aircraft, unless it could be done before take-off, it is far more difficult. Also, the need to use phase-comparison comes about mainly because long range at sea-level is wanted and the only radio frequencies that can provide this are in the medium or low frequency bands (30–3000 kHz) where sharp pulses cannot be used. Aircraft can use their altitude to extend the range of line-of-sight systems to something reasonable and do not have to use low frequencies, unless like many helicopter operations they are confined to low levels.

GPS is not generally thought of as a continuous-wave phase-comparison system, but its signals are usable in a similar way. Its basic transmissions on the L1 (1575.42 MHz) and L2 (1227.6 MHz) frequencies are derived from a common oscillator and are phase coherent, being respectively the 154th and 120th harmonic of a 10.23 MHz reference oscillator. What this means is that their timings when they leave the satellite are exactly the same, a fundamental requirement for permitting both phase comparison and ambiguity resolution. If in addition to the standard method of code tracking a GPS receiver can also use cycle and phase-comparison techniques an astonishing increase of accuracy results, from the standard C/A (non-S/A'd) fifty metres to a few centimetres, due to the very short wavelength of a 1575 MHz signal – about nineteen centimetres. This technique is described in more detail in Chapter 3.

Appendix 2

How Satellite Positions In Space Are Defined

1. Cartesian Co-ordinates

Latitude and longitude are basically unsuited for defining positions in space because they are neither geocentric nor three-dimensional. What this means is that the centre of the spheroid on which they are based may vary depending on where the map was drawn and is in any case unlikely to be the same as the centre of the satellite spheroid. Also, lat/lon are tied to the Earth and rotate with it.

Instead, a right-angled Cartesian system centred on the mass centre of the Earth can be adopted (Fig A4). While this is computationally simple (and is the prime system used in GLONASS and Tsicada), it does not lead to any instinctive appreciation of the size, shape or orientation of an orbit. Nor does it degrade gracefully – if a navigation satellite's ground uplink fails, a Cartesian system will result in rapidly increasing errors, because it can only be projected using instantaneous velocity vectors.

The definitions of the various axes are:

'X' axis:
Projected outwards from the centre of the ellipsoid to the intersection of the zero meridian with the Equator and beyond.

'Y' axis:
Projected outwards from the centre of the ellipsoid to the intersection of the 90° E meridian with the Equator and beyond.

'Z' axis:
Projected outwards from the centre of the ellipsoid through the North Pole and beyond.

Note: The terms 'ellipsoid', 'zero meridian', 'Equator', and 'North Pole' are themselves the subjects of further definition but the conventional meanings are adequate here.

2. Keplerian Parameters

The name is derived from the famous Austrian Johannes Kepler, the first to describe the planetary Laws of Motion which state that a body in motion about the Earth will describe an ellipse with one of its foci at the mass centre of the Earth (Fig A5). A perfect circle is very rarely achieved because the Earth itself is not a perfect sphere and even if the satellite was launched into a perfect circle it would soon degenerate

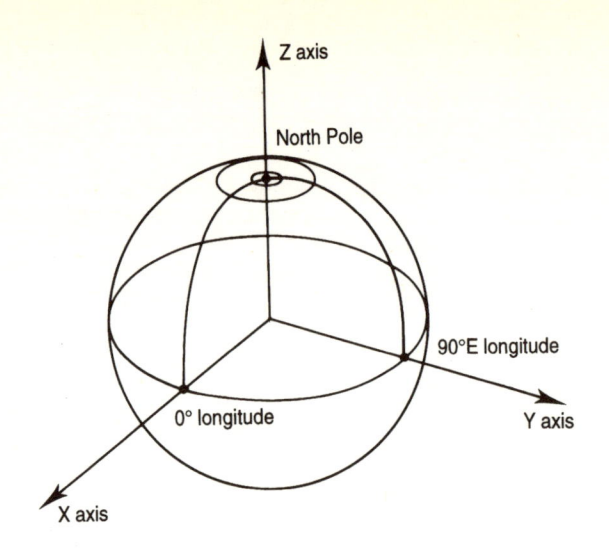

The Air Pilot's Guide to Satellite Positioning Systems

Fig A4 – The Cartesian System

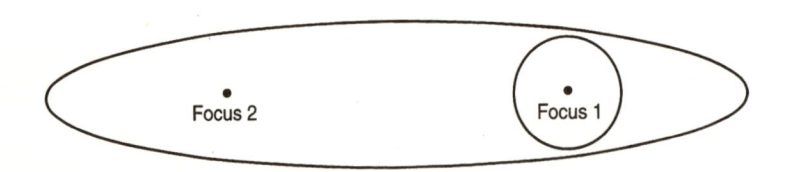

Fig A5 – Orbital Ellipse

into an ellipse. Satellites in *low* Earth orbit actually oscillate regularly between elliptical and circular orbits.

The shape of this ellipse is described by its ellipticity – the amount by which it departs from a perfect circle. A circle has an ellipticity of zero; most low Earth orbiters have ellipticities around 0.00001. Some satellites are deliberately put into very elliptical orbits, which have special properties, but so far they have not been used for navigational purposes.

The size of the ellipse is given as the radius of the smallest circle that can be drawn encompassing the ellipse at its widest part – its semi-major axis.

Between them, the ellipticity and semi-major axis provide all we need to know about the size and shape of a satellite orbit, but they tell us nothing about where the satellite is around it, nor how the ellipse lies relative to the Earth or space.

To provide these, we have to adopt reference points or planes. The first is the Equator, which of course is at right angles to the Earth's spin axis through the North and South poles. Like the satellite in its orbit, the Earth is spinning like a gyroscope in space, so the angle between the planes of the Equator and the orbital ellipse will be quite stable, and is called the inclination. It is a useful angle to know since it tells us how far North and South of the Equator a satellite will pass overhead. Thus, a satel-

Appendix 2

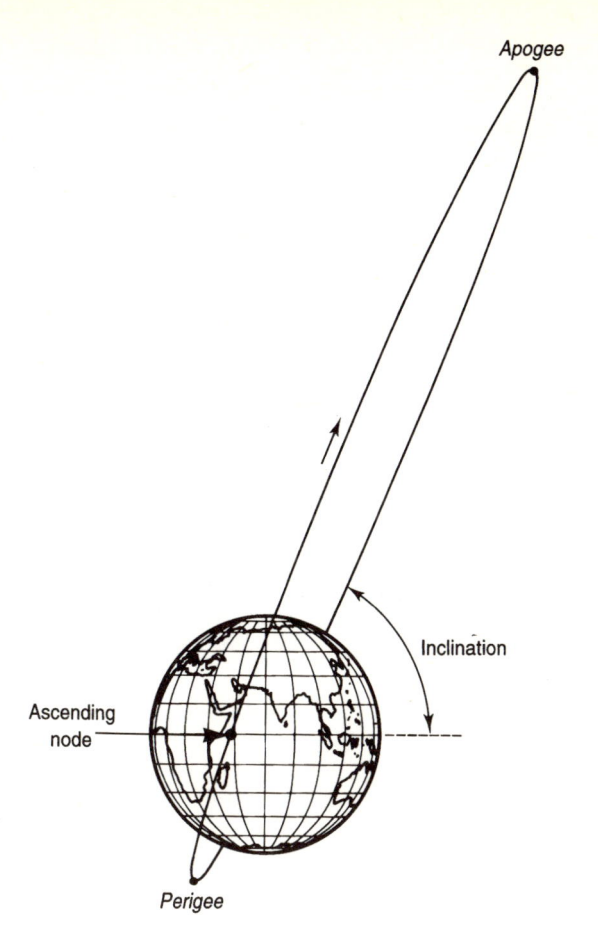

Fig A6 – Perigee and Ascending Nodes

lite with an inclination of 90° would pass over both Poles, and cover the entire Earth every twenty-four hours, which is why this inclination was used for the US Transit satellite navigation system, and for some meteorological satellites.

But we still do not know where the satellite is around its orbit, so now we must adopt two more reference points (Fig A6).

The first is the point round the orbit at which the satellite is closest to the Earth, its perigee, which is chosen because it is reasonably stable. Although it does in fact precess around the orbit because of the Earth's lack of sphericity, it does so relatively slowly, and if the inclination is 63.4° will not do so at all because at that inclination the two major forces causing precession cancel out. Satellite position round the orbit is therefore measured from this point and the time it passes through it is often used as the orbit reference time, although not always.

The second is needed because the Earth is spinning within the orbital plane and we

The Air Pilot's Guide to Satellite Positioning Systems

have to reference the orbit to the Earth. The point where the orbital plane crosses the Equator at the reference time, the 'ascending node', is compared either with the Greenwich meridian, when it becomes the longitude of the ascending node, LAN, or a reference point in space known as the vernal equinox, or the 'first point of Aries', (the point where the Sun crosses the Equator northbound in the Spring), when it is called the Right Ascension of the Ascending Node (RAAN).

Definitions

APOGEE The point in its orbit when a satellite is furthest away from the Earth.
PERIGEE The point in its orbit when a satellite is closest to the Earth.
ARGUMENT OF PERIGEE (1) The angular distance, measured in the orbital plane, in the direction of motion of the satellite, from the point of intersection of the orbit and equatorial planes, to the perigee point.
Satellites move faster at Perigee than Apogee and this information is necessary to accurately calculate their position. If the Argument of Perigee is 0° then the lowest point of the orbit of that satellite would be at the same location as the point where it crossed the Equator in its ascending node travelling South to North. If the argument of Perigee is 180° then the lowest point of the orbit would be on the Equator on the opposite side of the earth from the ascending node when it would be travelling North to South.
ASCENDING NODE Point at which the satellite crosses the Equatorial plane travelling South to North. (See also Right Ascension Of The Ascending Node.)
ECCENTRICITY (2) A unitless number which describes the shape of the orbit in terms of how close to a perfect circle it is. This number is given in the range of 0.0 to just less than 1.0. A perfectly circular orbit has an eccentricity of 0.0. A number greater than 0.0 represents an elliptical orbit with an increasingly elliptical shape as the value approaches 1.0.
EPHEMERIS A tabulation of a series of points which define the position and motion of a satellite.
EPOCH (3) A specific time and date which is used as a point of reference; the reference time for which an element set for a satellite is valid.
EQUATORIAL PLANE An imaginary plane running through the centre of the earth and the Earth's Equator.
INCLINATION (4) The inclination of a satellite refers to the angle formed by the orbital plane and the equatorial plane. The inclination is always a positive number of degrees between 0° and 180°. Zero angle of inclination indicates a satellite moving from West to East directly along the Equator. All geostationary satellites have inclinations close to zero. An inclination of 28° (most Shuttle launches) indicates an angle of 28° between the equator and the orbit of the satellite. Also, that satellite will travel only as far North and South as ± 28° latitude. On its ascending orbital crossing of the Equator (moving from South to North), the satellite will actually be moving from South-west to North-east. An inclination of 90° means that the satellite moves from South to North crossing directly over the North and South poles. A satellite with an inclination greater than 90° is said to be in a retrograde orbit, meaning that the satellite moves in a direction opposite to the rotation of the earth. A satellite with an incli-

Appendix 2

nation of 152° moves from South-east to North-west as it crosses the Equator from South to North. This satellite will move as far North and South of the Equator as 28° latitude but travel in an orbital direction exactly opposite a satellite with an inclination of 28°.

MEAN ANOMALY (5) Angular distance measured in the orbital plane, in the direction of motion, from the perigee point to the satellite's position at epoch time. The Mean Anomaly fixes the position of the satellite around its orbit. An Argument of Perigee of 0° would place perigee at the same location as the Ascending node. If in this case the Mean Anomaly were also 0° then the satellite's position would be located directly over the Equator at the ascending node. If the Argument of Perigee was 0° and the Mean Anomaly was 180° then the satellite's position would have been on the other side of the earth over the Equator but travelling North to South.

MEAN MOTION The number of complete revolutions a satellite makes in a given unit of time, usually measured in revolutions per day (24 hours, 1,440 minutes, 86,400 seconds).

RIGHT ASCENSION OF THE ASCENDING NODE (6) The angular distance from the Vernal Equinox (see below) measured eastward in the equatorial plane to the point of intersection of the orbit plane where the satellite crosses the equatorial plane from South to North.

SEMI-MAJOR AXIS (7) Because the orbital shape is an ellipse there are two axes at right angles to each other. The smaller is called the minor axis; the larger the major axis. Only half the axes are usually specified because if one of the elliptical foci is close to the centre of the Earth the semi-major axis will be very close to its actual distance from the centre of the Earth. Subtracting the radius of the Earth gives the height of the satellite above the Earth's surface, important for estimating its elevation and azimuth from a geographical point.

VERNAL EQUINOX Also known as the first point of Aries, being the point where the Sun crosses the Earth's Equator going from South to North in the spring. This point in space is essentially fixed and is therefore used as the reference axis.

LONGITUDE OF ASCENDING NODE This is similar to RAAN but instead of measuring from the vernal equinox the longitude of Greenwich (0) is used. This is used in the GPS system and has the advantage that the position of the Vernal Equinox (the 'first point of Aries') does not need to be calculated first, thus simplifying calculation in the receiver.

The seven parameters numbered 1–7 above are all that are needed to be able to determine a satellite's position at any time. The sharp-eyed might object that although they describe the geometrical shape and size of the orbit they don't say anything about how fast the satellite is travelling or how many revolutions it makes each day. The fact is that for a given semi-major axis both of these are fixed and cannot be varied. Application of power to a satellite by firing its motor does not change its speed while remaining at a constant height but instead drives it into a higher orbit. A bit like an aeroplane flying at a constant angle of attack – really! So given the semi-major axis orbital speed can be calculated – and the other way round. Actually, the easiest way of measuring satellite height is simply to measure how long it takes for one orbit.

These parameters produce an idealised smooth orbit but it is rarely the case in practice that the actual orbit is so smooth. There are perturbing factors other than the Earth's gravity itself, such as the gravity field of the Moon, radiation pressure from the Sun, and so on. Navigationally they are important disturbing influences, so a mod-

The Air Pilot's Guide to Satellite Positioning Systems

ified form of Keplerian parameters known as 'pseudo-Keplers' are used. They look exactly like ordinary Keplers and can be treated mathematically the same way, but are actually 'best fit' data valid only over a small part of the orbit and are calculated to give best results only when the satellite is in that part of its orbit. Because they are only valid for small parts of each orbit, they would cause considerable errors if they were used for forecasting satellite positions many days or weeks ahead, so it is common practice for GPS and other systems to transmit an additional set of true Keplers for each satellite to be used only for forecasting positions. This set is usually called the 'Almanac'.

Appendix 3

The Transit System

The Doppler/Fizeau principle on which Transit depends is well-known and is used for many types of measurement. A change of frequency occurs at a receiving site whenever the distance between it and a transmitter is changing. It occurs with all types of acoustic or electromagnetic wave, hence the usual illustration of car sirens changing in pitch as they pass. Its magnitude depends on the *rate of change* of distance and the actual wavelength in use. It follows that the most accurate measurements will be made when the wavelength is very short (very high frequency). It is of course a requirement that the transmitted frequency does not itself change over the period of measurement and also, less obviously, that the reference frequency in the receiver does not either.

If it can be assumed that neither drift significantly, then a series of Doppler-shift measurements taken over a period of a few minutes can be translated into slant ranges that are unique to a particular point on Earth (Fig A7). The measurements are made quite simply by comparing the frequency at which the satellite transmission is received with a stable reference frequency in the receiver.

The system provides a reasonably accurate fix each time a satellite pass occurs, but is of course of no value in between passes. Since its major purpose is the updating of

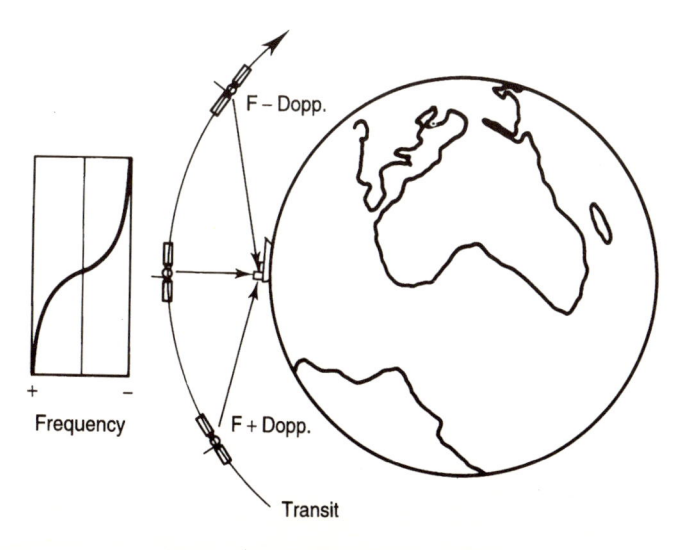

Fig A7 – Doppler Principle in Transit

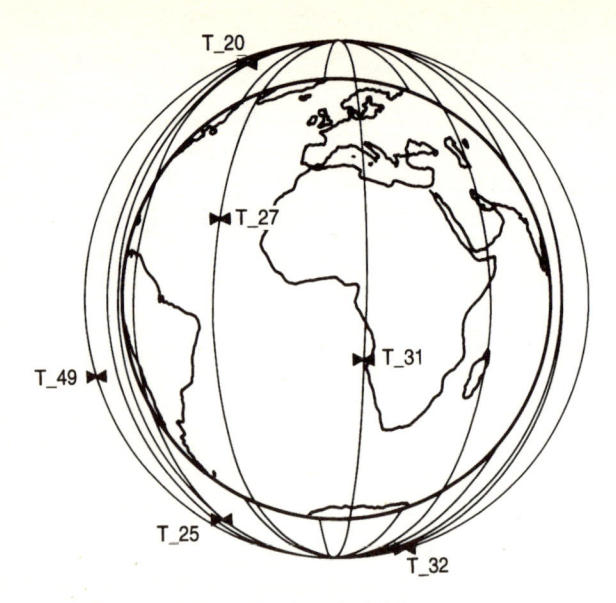

Fig. A8 – Transit Orbits

marine inertial systems (it is operated by the US Navy) this is of little consequence, but it makes it unusable in aircraft that might travel over 1,000 miles between updates, even if the Doppler-shift caused by the aircraft itself did not cause any errors. Over the years Transit has been very considerably refined and has provided the major means whereby the evaluation of the precise shape and size of the Earth itself has proceeded. Retrospectively it might even be considered that its major contribution has been to the geodesy rather than to navigation. Due for closure in 1996, it is now operating with twelve satellites in orbit, although only a maximum of eight are switched on at any one time because of system limitations.

Operation

Transit satellites orbit the Earth in polar orbits at a height of about 1,000 km and with a period of about 1hr 47m (Fig A8).

At any one place a satellite will be above the horizon for about sixteeen minutes per orbit. At the exact instant the satellite is at minimum range, there will be no Doppler offset (the range is momentarily not changing). At this point the receiver must be on a circle of position lying at right angles to the satellite orbital plane. The geometric characteristics of the orbital plane are known, and the satellite's position along it at any time (from the broadcast message), therefore one line of position can be obtained. The second – actual range from the satellite at this time – is obtained by measurement of the rate of change of frequency near the mid-point. Transit uses two radio frequencies – 150 and 400 MHz – to be able to measure ionospheric refraction. At an orbital height of 960 km Doppler offsets will be 3.7 and 9.8 kHz respectively as the satellite comes up over the horizon on a pass that will take it overhead.

Appendix 3

The satellites are travelling at a rate of 7.37 km/s so a timing precision within milliseconds is needed if position in orbit is not to be wrong. Without assistance it is unlikely that receivers could maintain this accuracy for very long, so it is arranged that the satellites transmit two-minute time markers to which the receiver locks and corrects its own timing. Many low-cost receivers only measure Doppler over these two-minute blocks, but the more accurate survey versions often measure Doppler over intervals as short as 4.6 seconds (the data word length).

Orbital data is continuously transmitted as part of the message, in the form of Keplerian parameters, enabling the receiver to calculate satellite position at any desired time. The message is composed in such a way that accurate data is available for two minutes ahead of and behind the current time, so at any time the receiver has three sets of similar data available to it. Usually it 'majority-votes' these to produce one set of high integrity. The satellite gets this information (and its timing marks) from a ground upload station that in turn gets it from a central computer. This computer uses tracking data from a tracking network to calculate satellite orbital parameters and then predicts what they will be for some twelve to eighteen hours ahead. The satellite transmits these and discards old information as it orbits so that it is always transmitting current data. After eighteen hours or so its memory is emptied unless it is uploaded again.

The ground tracking network has four tracking stations located in the USA, the computing centre being in California. There is an auxiliary tracking network using twenty-two stations around the world that provides data for the computation of precise orbits calculated some days later and available for those geodesists needing very accurate data.

Accuracy

Single-channel receivers using the 400 MHz signal only are unable to correct for ionospheric refraction and may suffer up to 250 metres error from this source. Atmospheric refraction may contribute another 20–30 metres. There are other variable errors depending on elevation angle at nearest approach, the best elevations being between 30° and 50°. Satellite passes at lower than 10° and above 75° are not really usable and most receivers make provision in their software to gate them out.

Dual-channel receivers make an almost complete correction for ionospheric refraction but are just as liable to atmospheric refraction and the elevation limits. This class of receiver usually contains a better frequency reference than the simpler single-channel units and is more accurate all-round.

Both types suffer from the inherent problem in the system that any motion on the part of the user will cause its own Doppler-shift that will be added to or subtracted from that of the satellite and hence will cause an error. If the vessel is sailing due North or South this Doppler will cause maximum error (the satellites are also travelling North/South), but if East/West the effect is small. At its worst, 1 knot of unknown speed might cause up to 400 metres fix error. It can be corrected if velocity is exactly known (usually from some other system), but it is another of the reasons why aircraft have never used it.

Transit Satellites in Orbit

Between 1985–1987 nearly all remaining usable Transit satellites were launched into orbit two at a time in the SOOS program (Stacked Oscar On Scout) to be switched on as and when required. The ground computer cannot handle more than eight, so only seven or eight are actually transmitting at any one time although there are twelve in orbit and usable. The eight are selected on the basis of accuracy and even spacing of passes. Three satellites are of the much improved Nova type that are self-stabilising in orbit and thus are always kept on, whereas all the others are of standard Oscar design and are switched as required.

Future of Transit

The US Navy, who own the system, intend to switch it off in 1996 consequent on the successful introduction of GPS. They have stated that it will not be handed over to any civil agency or commercial organisation.

Appendix 4

Spheroids and Datums

1. Spheroids and Ellipsoids

The terms 'spheroid' and 'ellipsoid' are for all practical purposes indistinguishable when applied as descriptive terms to the shape of the Earth. What they refer to is the fact that the Earth is not a sphere although extremely close to one. Newton showed as long ago as the 17th Century that the Earth had to be a flattened sphere in order to maintain equilibrium, its rotation causing fluids such as water, air and its own liquid centre to migrate towards the Equator. This causes it to be slightly flattened at the Poles and become an ellipsoid but the amount of flattening is tiny, about 1/300. If a circle of radius 6 cms (about 2.5 inches) is drawn to represent the Earth, this flattening is about 0.2 mm on the same scale, or less than the thickness of a line. Nonetheless, it is important at the accuracies available from satnav and must be allowed for.

The size and shape of the spheroid is given by two parameters, semi-major axis and flattening or ellipticity. They are very similar to the two Keplerian elements semi-major axis and eccentricity used to describe satellite orbits.

Table A1 – Spheroids In Use In Europe

Name	Date	Semi-major Axis	Flattening	Use
Airy	1849	6,377,563	1/299.32	UK
Clarke	1880	6,378,249	1/293.47	France
Struve	1860	6,378,298	1/294.7	Spain
Hayford or Int'national	1909 1924	6,378,388	1/297.00	World
GPS spheroid: WGS-84	1984	6,378,137	1/298	GPS

Different values were arrived at because they were determined mainly in the 19th Century by astro-geodetic measurements originating in different parts of the World. They fit observations made in their 'home' territories quite well but not necessarily those made elsewhere.

By definition spheroids are regular mathematical figures and take no account of local variations in the shape of the Earth. Satellites revolutionised the measurement of more localised variations and led to the modern concept of the Earth as being pear-shaped with a number of bumps and hollows. This more accurate figure of the Earth is referred to as the Geoid.

The Air Pilot's Guide to Satellite Positioning Systems

To give a sense of proportion, while the difference of any of the spheroids from a perfect sphere can be as much as 11,500 metres, the maximum difference of the refined geoid from a spheroid is only about 80 metres. Their importance in fixing position is that very approximately a 1-metre error in the vertical direction is reflected into a 1-metre error in the horizontal plane, so if the difference between the geoid and the spheroid were ignored then in some areas of the World an 80-metre error might be introduced into the horizontal fix component irrespective of the accuracy of anything else.

2. Datums

Knowing the exact shape and size of the Earth is one thing; defining position on it is something else. If on land we can always use the nearest identifiable object, but if above cloud, out at sea, or in the desert this might be a little difficult. The universal method is to put down an artificial grid and relate everything to that, the best known being latitude and longitude. There are plenty of others; in the UK there is the National Grid which appears on all Ordnance Survey maps, and offshore surveyors often use a UTM (Universal Transverse Mercator) grid.

They all have one thing in common – they must start from somewhere, and this is their datum. Lat/lon is world-wide and uses the position of the Poles but other grids are more local and may use some datum nearer home. The National Grid uses a position just to the South-west of Land's End, ensuring that all grid positions in the UK are to the North and East.

There is another complication in that these theoretical grids have to be related to the real world. Before satellites it was done by first establishing a baseline somewhere convenient, measuring it very accurately, and triangulating outwards from it until the area of interest had been covered. This was often an entire country and might entail considerable distances being covered, so the accuracy of the survey at its boundaries was unlikely to be as accurate as at its baseline. Then, the theoretical grid was tied to the resultant real–life network by adopting some fixed point as the basic reference origin. Most countries adopted the position of their national observatory which had the necessary equipment to make astronomical observations.

The net result was a plethora of national systems unrelated to each other, all on different datums and using different spheroids. Little wonder that maps originating in different countries did not match at boundaries.

The first serious attempts at common datums and spheroids did not occur until after World War 2, when the European Datum of 1950 (ED50) was surveyed. This is centred at Potsdam, Germany and extends over the whole of Europe.

The situation in the UK is described in more detail in Appendix 5 following.

Table A2 – European Mapping Datums

Name	Date	Origin	Offsets from WGS-84 (satellite) (feet).		
			X	Y	Z
ED50	1950	Potsdam	–84	–103	–127
OSGB36	1936	Greenwich	368	–120	425

Appendix 5

Ordnance Survey Maps and Mapping Policy

Acknowledgement is made to the Ordnance Survey for the following information.

1. The 'WGS-84 Problem' in the UK

In Britain the National Mapping Agency, Ordnance Survey, uses a National Grid as the basis of almost all map referencing. It is a rectangular co-ordinate system allowing users to reference any position within Britain, but being a 'local' system based on a particular representation of the curved earth on to a flat surface cannot be extended to Ireland, France or anywhere outside Great Britain.

The framework on which Ordnance Survey maps are created is:
* Definition of the most appropriate geodetic datum (i.e., shape, size, orientation and position of the Earth) for mapping Great Britain;
* Triangulation to provide the datum and control for the mapping;
* A map projection to project the spherical Earth on to a flat map;
* A co-ordinate system to provide references on the map – the National Grid.

They are closely linked. The shape of the Earth itself is approximated by a spheroid – the closest regular mathematical figure to the true shape of the Earth over the UK. The spheroid used in the UK for OS mapping is known as a 'modified Airy Spheroid'.

The Primary Triangulation of Great Britain (involving triangle sides of 50 to 70km) defines primary geographical control points which are used to control more detailed surveys. The positions of these points are given in latitude and longitude based on the UK geodetic datum known as Ordnance Survey Great Britain 1936 (OSGB36).

This primary control physically exists on the curved surface of the Earth and the result is that when features so mapped are transferred to a flat map there is distortion – it is like trying to flatten out a segment of a ball without splitting it. Many mathematical transformations, called 'projections', have been devised in order to minimise one or another of these distortions, the well-known Mercator being one of them. The UK Ordnance Survey has standardised on the Transverse Mercator projection for this purpose. Having produced a flat map, distortions and all, it is more convenient to have a simple rectangular grid overlaid on it purely for the purpose of referencing features, rather than use a curved and distorted true geographical grid like lat/lon. The National Grid has been adopted for this purpose in the UK and is a rectangular co-ordinate system designed so that scale errors, particularly at the edges (e.g., in the Hebrides), are minimised and co-ordinate values are always positive. Lines of latitude and longitude are curved on OS maps but the conversion between National Grid co-ordinates and latitude and longitude is well defined and mathematically straightforward.

2. Ordnance Survey Maps in the UK

Before 1936 all large-scale mapping in Britain was based on local projections (the 'County Series'). Each county or group of counties had a projection of its own but it proved difficult to reconcile map detail across county boundaries – especially when those boundaries changed – and eventually this led to the creation of OSGB36 and the National Grid. OSGB36 proved adequate for its time and has formed the basis for all OS mapping since.

During the last sixty years, however, geodesists have developed much better knowledge of the shape of the earth and the positions of control points used for mapping. This information has come from two advances. In the first place, new survey and computational methods allow much more accurate determination of position and more rigorous calculations. Secondly, navigation satellite data has forced a global – rather than a national or regional – view to be considered, requiring in turn a universal definition of the shape of the Earth. There has therefore been a gradual improvement of the scientific definition of position, but without much effect on practical map users until now. Many of the changes are minor and the improvement to absolute position is irrelevant to the many users of map information who are only interested in relative position (e.g., 'the gas main is two metres from the edge of the pavement').

3. The Effect of New Survey Methods

In 1950 a European datum (ED50) was defined primarily to assist positioning offshore structures and boundaries. In the 1970s and early 1980s, the survey information used to create OSGB36 was enhanced with distance measurements and satellite-derived positions and recomputed. The last of these scientific computations was the Ordnance Survey (Scientific Network) 1980 or OS(SN)80, which included data from the Ordnance Surveys of Northern Ireland and the Republic of Ireland. It produced a new geodetic datum but one designed purely for scientific rather than mapping purposes.

On a world level, new datums were also devised to accommodate positions both on the Earth's surface and above it – satellite orbits for example. The latest of these World Geodetic Systems (WGS) is WGS84 which supersedes WGS72. These datums are used with a geocentric (i.e., Earth-centred) co-ordinate system instead of spherical (latitude and longitude) co-ordinates.

During the late 1980s GPS started to be widely available. Since it is a satellite system with global coverage, the reference frame has to be global and GPS observations resulted in co-ordinates defined in terms of WGS84. It was realised that a more precise definition (to centimetres rather than the decimetres given by WGS84) was both required and possible. As a result, a campaign was begun to improve the definition of WGS84 in Europe which has now produced the European Terrestrial Reference Frame 1989 (ETRF89). This is now the fundamental geodetic reference framework throughout Europe.

What has become apparent is that the original OSGB36 triangulation and its equivalents in other countries are not directly compatible with ETRF89. Furthermore, because of the way in which they were calculated, the differences are not uniform – in other words a simple scaling, translation and rotation will not suffice. There are discontinuities both between national systems and within them. The net effect is that, in ETRF89 terms, the present National Grid is no longer rectilinear. This situation is not unique to Britain; there is little or no mapping anywhere at present which is based upon WGS84 or ETRF89. Considering this and that most map users in Britain are happy with the National Grid, why should they now have to consider the implications of new geodetic datums?

The main reason is that GPS receivers give users access for the first time to highly accurate, virtually instantaneous, global positioning information. However, because of the factors described above, co-ordinates in WGS84 differ from co-ordinates in OSGB36 by up to hundreds of metres. This is not an error; it is a consequence of the use of different geodetic datums.

4. National Grid and GPS Co-ordinate Compatibility

All national mapping agencies, not only OS, are now faced with a dilemma. Should they continue to provide mapping based on what are now known to be distorted models, or should they transform all existing mapping to a new co-ordinate system?

If existing co-ordinate systems are retained, mapping organisations may be building up potential problems for the future, but by doing so printed maps and existing data can be maintained as they are. Users could continue to use the system with which they are familiar and in which they have invested heavily.

However, there is an increasing amount of data becoming available that is global in extent and based on the new geocentric co-ordinate systems (e.g., satellite-based remotely sensed data) and requires conversion.

The advantages of the conversion of data lie in the long-term ability of users of GPS and other positioning systems to have a common spatial reference. Cross-border mapping will become much more feasible – a factor of particular importance in mainland Europe. The disadvantages are first in the cost involved and (more significantly in the short term) there would be side effects such as the alteration of map sheet boundaries. These effects would mean that users of the mapping who hold their own data based on existing OS National Grid co-ordinates or sheet lines would have to convert both datasets in sympathy.

Methods exist that can be used to improve compatibility between the National Grid and GPS co-ordinate systems to about five metres. Although adequate for some purposes, it is unsuitable for applications requiring precise positioning. As GPS techniques become more sophisticated, there is therefore an increasing need to have transformations that make OSGB36 (and hence the National Grid) and GPS derived co-ordinates as compatible as possible.

5. Reference Systems and Geodetic Transformations

The Ordnance Survey now accepts the need for it to take a lead on these aspects of geodesy and has defined a policy on how to deal with transformations between the various systems (primarily National Grid to ETRF89 or WGS84 and vice versa). The accuracy achieved by the various types of transformation will be within levels agreed to suit different applications.

Users of existing maps or digital map data in Great Britain who are using – or considering the use of – GPS should make themselves aware of the implications of using different geodetic datums and reference systems. Similarly, anyone setting up a Geographic Information System that will take from outside their parent organisation, for instance from OS, should take into account the possible need for conversion of the data in the longer term.

For the time being, OS has decided not convert its data or maps to a new reference system. However, this will be reviewed periodically; in the long term it is almost certain a change will be made. The OS emphasise that no-one should consider the existing system to be wrong. The National Grid system is an entirely safe and accurate system provided that users do not wish to integrate National Grid co-ordinates with GPS ones or to measure interpoint distances to accuracies of centimetres.

6. The UK Ordnance Survey Policy Statement on Reference Systems and Geodetic Transformations

Background
Some modern survey systems – notably GPS (the Global Positioning System) – provide positional co-ordinates for ground detail which cannot at present be related easily, accurately or consistently to the National Grid co-ordinates determined from OS map information or by traditional survey techniques based on monumented OS control points. At a public meeting held in London on 10 December 1993, those present called for national standards to be put in place so that survey and geographical information could be merged and managed without loss of accuracy, whatever referencing system had been used. In particular, a need was identified to be able to convert easily and accurately from National Grid (OSGB36) co-ordinates to ETRF89 co-ordinates (closely related to GPS-derived ones) and vice versa.

Ordnance Survey recognises the need to take a national responsibility and for it to lead in this area. A critical factor permitting this leadership is now in place; OS is currently finalising an OSGB36-ETRF89 transformation technique that will give an accuracy of better than 20cm over mainland Great Britain.

Ordnance Survey Policy
In response to opinion and to ensure that OS mapping and GPS-based data in particular can be used in conjunction without loss of accuracy or quality Ordnance Survey will:

* retain its mapping for the time being in the National Grid (OSGB36) referencing system;

Appendix 5

* periodically review the possibility of adopting ETRF89 or a related system as the basis for its mapping and spatial databases, taking into account customer demand and the views of the national GPS, GIS and mapping communities;
* make widely available, at no cost, a compact transformation for converting co-ordinates anywhere in Great Britain to and from OSGB36 to an accuracy of two metres. Representing an improvement on existing national transformations, this product will be particularly suitable for use within GPS equipment programmed to give direct OSGB36 positions and in computer software;
* provide a bulk transformation service for conversion of user datasets and the OS data they hold between different referencing systems. The accuracy involved will be chosen appropriate to customers' needs;
* provide a comprehensive geodetic transformation advisory and consultancy service. For example, OS will assist customers in assessing the feasibility of upgrading data that has already been converted using a low accuracy transformation. Where required, OS will also assist customers in implementing such schemes.

Definitions

ETRF89 – European Terrestrial Reference Frame 1989. A geodetic framework resulting from EUREF designed to give more precise co-ordinates throughout Europe. In practice it is compatible with WGS84 to within 1 metre.
EUREF – European Reference Frame. A sub-commission of the International Association of Geodesy responsible for the definition of a true tri-dimensional control network covering Europe. The definition was carried out using very high accuracy techniques, such as Very Long Baseline Interferometry which gives centimetric accuracy over thousands of kilometres.
GEODESY The science of the shape and size of the Earth together with its gravity field.
GEOID The equipotential surface of the Earth; the surface that would be taken up by water if all the oceans were connected together through mountain ranges and were unaffected by gravitational forces other than that of the Earth. Important because all observations taken using spirit-levels or gravity (sextants, theodolites) are referred to it.
GEODETIC DATUM A definition of the Earth's position, shape, size and orientation.
GRATICULE The representation of intersection of lines of latitude or longitude (see spheroid) on the flat surface of a map. Depending on the map projection these may be straight or curved in one or both dimensions.
GRID A rectangular network used for co-ordination, normally on a flat surface.
MAP PROJECTION The projection of the curved surface of a spheroid onto a flat surface for maps.
NATIONAL GRID A rectangular co-ordinate system used within Great Britain for mapping.
OSGB36 A Retriangulation carried out in Great Britain between 1936 and 1962 to provide a framework for OS mapping.

The Air Pilot's Guide to Satellite Positioning Systems

OS(SN)80 A recomputation of the primary triangulations of Great Britain, Northern Ireland and the Republic of Ireland incorporating up-to-date survey measurements and computational procedures. Not used for mapping.

SPHEROID A mathematical solid providing the best estimate of the average Earth's surface either as a whole or locally. It is an ellipse of rotation about the minor axis. Co-ordinates on the surface of the spheroid are normally given in latitude, longitude and height.

WGS84 A global geodetic system defining the geodetic datum used by GPS. The definition is based on physical properties of the Earth such as its Angular Velocity and Gravitational Constant.

Appendix 6

International Frequency Allocations for Navigation

1. The International Telecommunications Union (ITU)

The ITU is the international controlling body for all matters relating to radio systems. It exists by virtue of an agreement between nations to co-operate on telecommunications matters, entered into freely by most of the sovereign states of the world. Although signatories have agreed to respect the regulations and use the mechanisms of the ITU, they have not relinquished their sovereignty. Each country has reserved the right to do what is necessary to protect its own national interests.

2. ITU Requirements

ITU Radio Regulation 2020 requires that:

'No transmitting station may be established or operated by a private person or by any enterprise without a licence issued in an appropriate form and in conformity with the regulations by the government of the country to which the station in question is subject'.

Under this regulation each country itself determines whether or not a radio transmitter may operate in its territory. Transmitters may even be authorised to operate in contravention of the ITU Radio Regulations if a country so wishes, and that has in fact happened. If a country finds that its radio systems are being interfered with by those of another, its procedure is to file a formal complaint directly with the administration of the country concerned. The ITU has no powers to involve itself in matters of this sort.

This Radio Regulation has been universally adopted and therefore all that is *legally* required to operate a radio navaid is the appropriate licence to transmit, usually granted by the department of the Telecommunications Administration charged with the duty of issuing them.

The major matter of concern to these licensing authorities is the allocation of a radio frequency. To assist in this, the ITU has classified all possible applications of radio transmitters and assigned blocks of radio frequencies to each application. Because it is very difficult to ensure that no transmitter will ever interfere with another, classes of radio service are also defined where, at the top level, every effort is made to ensure non-interference, while at the lowest level interference must be accepted if it occurs.

155

3. ITU Definitions and Categories of Radio Determination Systems

All radio services are categorised and this determines whether it will be allocated Primary or Secondary use of a frequency. The following definitions are those given in the ITU's 'Radio Regulations'.

3.1. 'Radiodetermination'
'The determination of the position, velocity and/or other characteristics of an object, or the obtaining of information relating to these parameters, by means of the propagation properties of radio waves.'

3.2. 'Radiodetermination Service'
'A Radiocommunication service for the purpose of Radiodetermination.'

3.3. 'Radiodetermination Satellite Service'
'A Radiocommunication service for the purpose of Radiodetermination involving the use of one or more space stations.'

3.4. 'Radio Direction-Finding'
'Radiodetermination using the reception of radio waves for the purpose of determining the direction of a station or object.'

3.5. 'Radionavigation'
'Radiodetermination used for the purposes of navigation including obstruction warning.'

3.6. 'Radionavigation-Satellite Service'
'A Radiodetermination-satellite service used for the purpose of Radionavigation. This service may also include feeder links necessary for its operation.'

3.7. 'Radiolocation'
'Radiodetermination used for purposes other than those of Radionavigation.'
(For example, radar).
　　Another term often seen in connection with radiodetermination systems but not defined by the ITU is 'Radio-positioning'. This was formerly used primarily by the offshore industry to designate a radiodetermination service providing position and/or velocity of an object but not necessarily used for its navigation. Often it is used for retrospectively deriving the object's position and velocity. It is now coming into use for all types of non-navigational position determination services.

4. Classes of User

4.1. 'Primary'
This class has first choice of radio frequency and is fully protected against radio interference. No other class of user may operate in the same frequency band unless it can

Appendix 6

prove that it will cause no interference. The Radionavigation Service is classified as a safety-of-life service and is accorded Primary status.

4.2. 'Permitted'
The same as Primary except that it does not have prior choice of frequency.

4.3. 'Secondary'
Very rarely has exclusive frequencies and must avoid interference to any other service except other secondary services that may be allotted the same frequency band later. All radiopositioning and most radiolocation services are in this category.

5. CCIR Frequency Allocations

The Primary frequency allocations to the Radionavigation-Satellite Service are:

149.9 to 150.05 MHz, 399.9 to 400.05 MHz
Used by the USA's 'Transit' and the Russian 'Tsicada' satellite navigation systems.
1240 to 1260 MHz, 1559 to 1610 MHz
Used by GPS and GLONASS.
32 to 33 GHz
Allocated to 'inter-satellite navigation'. As far as is known it is not currently in use.
43.5 to 47 GHz; 66 to 71 GHz; 95 to 100 GHz; 190 to 200 GHz; 252 to 265 GHz
Not in use.

CCIR allocations are made in relatively broad bands of frequencies and it is left to the individual country to allocate exact frequencies within them. This is done on the basis of prior occupancy, distance from other transmitters, and so on. In most radio communications applications, the exact frequency is not a matter of prime importance, but in radio navaids it may be very important where for instance, an exact frequency relationship is needed for its correct operation (e.g. Decca, Loran-C, etc).

Provided that the transmitter can be shown to be properly designed and operated so as to conform to whichever set of technical specifications is appropriate for its class of service, usually set so as to minimise interference, and an acceptable frequency is available, a licence is usually granted.

It is left entirely to the navaid operating agency to decide whether a radio navaid is necessary, how it will operate, and where it should be. The same agency pays its installation and operating costs and ensures its correct operation.

6. Public and Private Radio Navaids

A distinction is drawn between those navaids deemed essential for public transport safety and those installed for private commercial purposes such as offshore oil exploration and development, fleet tracking, and so on.

The Air Pilot's Guide to Satellite Positioning Systems

6.1 Public Services
These are installed by a Government agency when they are recommended by international agreement for transport safety purposes, typified by aviation VOR/DME and ILS systems. Because they are safety-of-life they are allocated exclusive frequencies and are closely regulated. The need for close regulation and supervision has meant that they are expensive to operate and unattractive to a private enterprise commercially. In any case, it has always been considered that such services are not appropriate for direct commercial operation, although private companies may be involved in their design, construction and maintenance.

6.2 Private Services
Permitted where there is a restricted and usually specialised user base that requires a specific service not available from the public services. The accuracy and coverage of public services may be insufficient or they may not exist at all. Offshore oil exploration survey is an example. In many areas where this is carried out there are no public services available at all, and in those where there are, they often cannot provide the required accuracies. These services are run on a commercial basis and technical characteristics are subject to agreement between operator and client. They are not subject to the same stringent regulation as a public service and are not open to public use. They are usually described as radiopositioning services rather than radionavigation services.

Appendix 7

The Spread-Spectrum Modulation Method

Since this is not an engineering handbook the description given here is sufficient only to provide a broad appreciation of the navigational effects of what is happening. For fuller details an electronic engineering book should be consulted.

1. Radio Energy and Bandwidth

The simplest and most basic form of radio wave is a plain continuous carrier; that is, an oscillation at a constant radio frequency with a constant amplitude. Nothing varies, both amplitude and frequency are steady and unvarying. This type of radio wave does not convey any information apart from the fact that it is there. The simplest way of modulating it is to switch it off and on; if this is done in a particular pattern then the pattern can be used to represent a message, as in the familiar Morse code (Fig A9).

For example, Morse code uses a short burst followed by a longer one to denote the letter 'A'. Morse code can also be sent by shifting the frequency slightly while keeping amplitude constant, requiring a slightly more complicated receiver capable of detecting frequency shifts. As was discussed in Chapter 2, if the carrier is switched on and off very rapidly, followed by a comparatively long pause before doing it again, the 'pulse' of radio energy can travel outwards to a target and be reflected back again before another one occurs, enabling range measurement.

The pulse cannot be switched on and off again in zero time and a small range-measurement error occurs due to this non-infinite switching time (the 'rise-time'), and it is important in radar systems to make it small.

The transmitted bandwidth of an absolutely pure carrier wave is almost zero if it is truly sinusoidal and absolutely constant in frequency and amplitude. The corre-

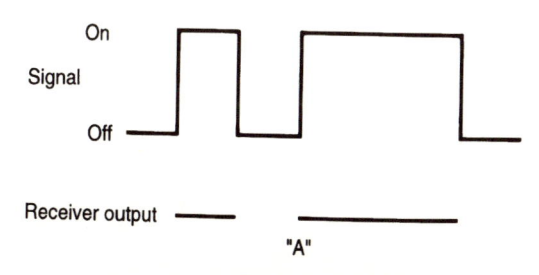

Fig. A9 – Morse Code Letter "A"

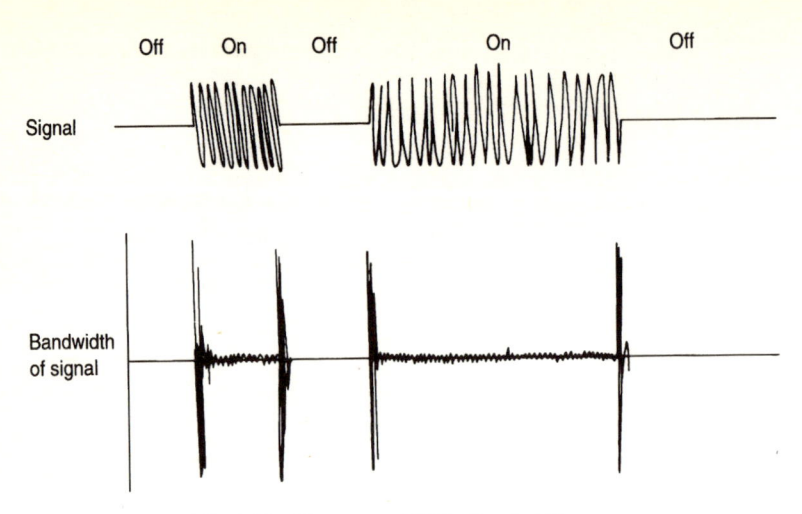

Fig A10 – Narrow and Wide Bandwidths

sponding receiver bandwidth needed to receive it can be made very narrow indeed, because the narrower it is the less radio noise is received although the carrier wave strength is unaltered. The Decca Navigator system is a good example of this. In normal operation it transmits only a carrier wave and the tracking bandwidth of a Decca receiver need only be a few Hertz, permitting the Decca transmitting frequencies to be only five Hertz apart. But when a carrier wave is switched on and off the energy required to do it appears as a wide band of radio-frequency energy whose bandwidth depends on the speed of switching. So at the instant of switch-on or off the bandwidth momentarily increases enormously although once the signal is established it decreases to near-zero (Fig A10).

Digital Modulation

In a pulsed radar system the delay between successive pulses can be used for signalling by systematically varying it in the same way as Morse code uses amplitude, which is called pulse-code modulation (PCM). This is an elementary form of digital switching, the modern form of which is to use microwave frequencies modulated by high-speed digital data streams producing extremely high signalling rates. This type of modulation is also known as 'spread-spectrum' because the original rather narrow-band radio signal is transformed into a wide-band signal by virtue of the high-speed switching it has undergone. This transformation has the effect that because the radio energy is now spread out into a wide bandwidth it is (at least in GPS) below the radio noise level as far as the usual type of receiver is concerned, unless a very high–gain aerial is used that can recover signal power without adding noise.

Appendix 7

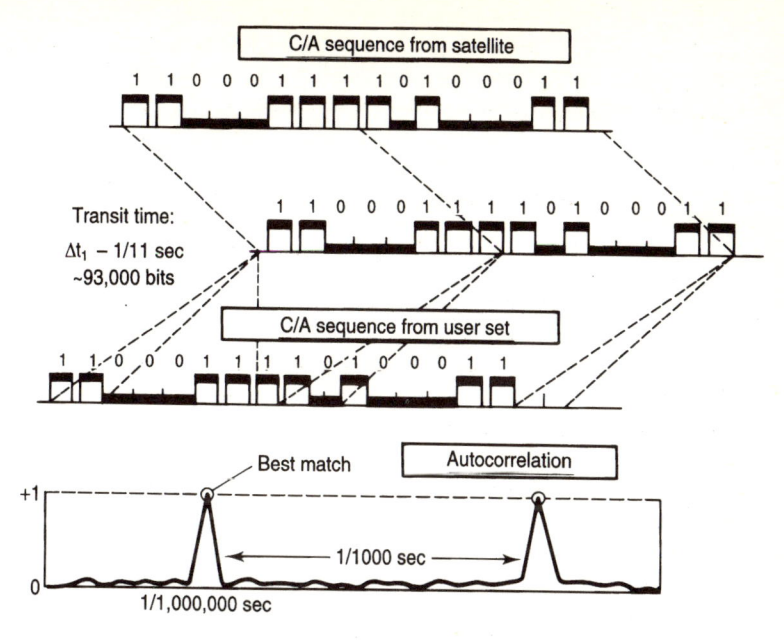

Fig A11 – Digital Demodulation

Spread-spectrum Demodulation

A receiver that does not know when the digital switching transitions occur cannot recover the original signal, but imagine that the receiver *already knows* the timing of the data-stream ('synchronous' operation). It can then generate the switching points of the data itself without requiring any input radio signal and can use them to decide when it should look for the incoming data. It can use a narrow-band filter for this purpose thereby getting rid of a lot of radio noise, improving the signal/noise ratio considerably and becoming able to detect a much weaker signal. This may seem rather a silly thing to do – if the data is already known why bother to send it? That would be true if it was the high-speed data that was actually wanted, but in GPS it is not – it is the absolute *timing* of when that data arrives and the 'data' in this case is the modulating bitstream. This bitstream can easily be arranged to have a unique 'word' – a special pattern – of bits that occur only at time intervals equal to or greater than the maximum range expected. The receiver's internally-generated replica of this unique word is then jumped one bit at a time until it fits the incoming bitstream exactly when it will have achieved time synchronism. It will know when it has done this because whenever its own bits occur there will be a corresponding incoming bit of radio energy and it can detect the systematic appearance of this energy in place of the random noise it would otherwise experience.

The time delay between its internal clock and the timing of its now GPS-synchronised internal bitstream is what constitutes the pseudo-range measurement, and produces the same effect as if the satellite had sent a single very short timing pulse. The bit length can be selected so as to give the required precison of measurement and in the case of GPS it is equivalent to 1 µS for the C/A code and 0.1 µS for the P-code. 1 µS is 300 metres of radio wave travel but the time of arrival of the start of the bit can be measured much more accurately than that, usually to better than one-tenth the bit length.

This technique has a number of advantages – it enables precise timing to be done; it needs only a low-power signal from the satellite; and because the timing word can be a different pattern for each satellite there can be several satellites on the same radio-frequency without any danger of confusing them. The additional RF energy on the frequency from the unwanted satellites is ignored and becomes part of the general background noise, although if there are many satellites present then it does in fact very slightly increase the level of this noise.

The PR bitstream does not carry any system information and the satellite data message is carried separately. This is a comparatively low-speed data stream compared to the PR code (50 vs. 1,000,000 bits/sec) and can be carried on the satellite signal by phase-modulating it without significantly increasing its bandwidth. Because it is so long it is made up of many thousands of the PR bits and can be reconstituted without difficulty.

Signal Security

There are a number of advantages to the spread-spectrum method other than accurate timing. The modulating bit pattern may be encoded in many different ways and changed regularly, provided the receiver knows which pattern is being used. If the recurring pattern is very short and repeated rapidly, it is easy to detect and in fact the C/A signal uses a code that lasts only one millisecond, allowing the receiver to acquire the signal quite rapidly.

The P signal uses a much longer code which is changed every week and is almost impossible to acquire unaided. P code receivers first acquire the C/A code which then gives the necessary information (hand-over word HOW) to enable rapid acquisition of the P code. In addition, this code may itself be encoded and require a special module to decode. When encrypted in this way the P code becomes the Y code, and because the decrypting modules are not authorised for civil use P code cannot be used by civil users.

Another benefit is that narrow-band interfering signals within the GPS band undergo a reverse process when the GPS signal is de-spread and their energy is spread out over the GPS band. This can sometimes help to limit interference effects.

Fig A12 shows how these various bitstreams are modulated onto a single carrier wave.

Measurement Accuracy

Early estimates of GPS ranging accuracy were that C/A code accuracy, since it used only one-tenth the chipping (pulsing) rate of P-code, would correspondingly be only one-tenth that of P-code. The connection is that a fast chipping rate permits, in effect,

Appendix 7

Combining the GPS signals

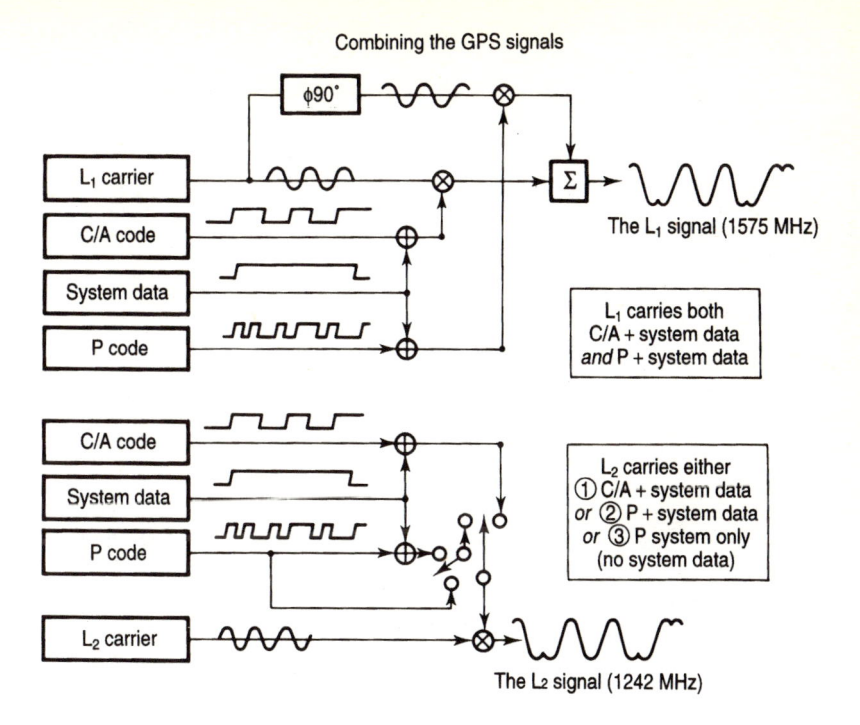

Fig A12

a faster rise-time for each bit that can be measured more accurately. In fact it turned out to be only two to three times worse and this was one of the reasons for the introduction of the Selective Availability feature. Recent work using modern estimation theory has confirmed that this reduced difference is to be expected and has shown how even more accuracy can be extracted from the C/A signal. It is now expected that the ultimate ranging accuracy of the C/A code signal, if governed entirely by what the satellite transmits and not by receiver characteristics, will be about twenty centimetres.

This raises the interesting possibility that since one wavelength at L1 is about nineteen centimetres it may become possible to use the code timing directly to resolve the nineteen centimetres ambiguities inherent in phase tracking techniques and thus increase overall ranging accuracy to a few centimetres in real time. This can actually be achieved with current (1995) survey receivers but depends on time being available (some few seconds) for averaging to be done, which restricts it to static or very slowly moving receivers. Newer techniques are under development which may permit its use in aircraft.

A ranging accuracy of a few centimetres achievable in real-time would result in a fixing accuracy of better than half a metre which would have important implications for precision approach techniques if propagation-related errors could also be reduced.

Appendix 8

Extract from the US 'Federal Radionavigation Plan, 1992'.

The Global Positioning System, GPS, Types of Service

GPS provides two services for position determination, the Standard Positioning Service (SPS) and the Precise Positioning Service (PPS).

1. Standard Positioning Service (SPS)
SPS provides a standard specified level of positioning, velocity, and timing accuracy available without qualification or restriction to any user on a continuous worldwide basis. The accuracy of this service is established by the US Department of Defense based on US security interests. When GPS is declared operational, SPS will provide, on a daily basis at any position world-wide, horizontal positioning accuracy within 100 metres 2 drms and 300 metres 99.99 percent probability.

2. Precise Positioning Service (PPS)
PPS is the most accurate positioning, velocity, and timing information continuously available, world-wide, from the basic GPS. This service will be limited to authorized US and allied Federal Government and military users and to those civil users who can satisfy US requirements. Unauthorized users will be denied access to PPS through the use of cryptography. P code-capable military user equipment will provide a predictable positioning accuracy of at least 17.8 metres (2 drms) horizontally and 27.7 metres (2 sigma) vertically. Timing/time interval accuracy will be within 100 nanoseconds (1 sigma).
These accuracies are dependent upon:

- the number of satellites in the GPS constellation
- the orbits chosen for the satellites
- the location of the user
- the local visibility contraints on receiving signals from satellites
- the criteria for selecting four satellites from among the visible ones
- the magnitude of all the range measurement uncertainties, including the GPS receiver itself, experienced by users (User Range Error, URE).

They assume the user will use whichever four-satellite combination minimizes three-dimensional position DOP (PDOP). Accuracy projections are based upon a fully operational system: twenty-one (or more) healthy satellites, normal uploads by the Control Segment, etc. Satellite visibility depends upon local conditions. Sometimes it is possible to track satellites less than five degrees above the horizon,

Appendix 8

Table A3
GPS Fix Accuracies
(Accuracy)

Predictable	Repeatable	Relative
PPS		
Horz – 17.8 m	Horz – 17.8 m	Horz – 7.6 m
Vert – 27.7 m	Vert – 27.7 m	Vert – 11.7 m
Time – 90 ns		
SPS		
Horz – 100 m	Horz – 100 m	Horz – 28.4 m
Vert – 140 m	Vert – 140 m	Vert – 44.5 m
Time – 175 ns		

particularly while in flight, but close to touch-down it may not be safe to assume this. These accuracy simulations use five degrees.

Coverage
A 24-Block II satellite constellation (21 plus 3 spares) will provide world-wide three-dimensional coverage.

Reliability
GPS operational (Block II) satellites have a design life of 7.5 years. Reliability figures can only be determined after satellites are launched and data are collected and evaluated. With the planned replenishment strategy, a constellation of 21 satellites plus 3 operational orbital spares will provide a 98 per cent probability of having 21 or more satellites operational at any time.

Fix Rate
Availability of fixes from the system is essentially continuous, but some receiver designs may limit fix rate to less than theoretically achievable.

Capacity
Since the user does not have to transmit, capacity is unlimited.

Ambiguity
There is no ambiguity.

Integrity
GPS satellites are monitored more than 95 per cent of the time by a network of five monitoring stations spread around the world. The information collected by the mon-

The Air Pilot's Guide to Satellite Positioning Systems

itoring stations is processed by the Master Control Station at Colorado Springs, Colorado, and used to periodically update the navigation message (including a health message) transmitted by each satellite. The satellite health message, which is not changed between satellite navigation message updates, is transmitted as part of the GPS navigation message for reception by both PPS and SPS users. Additionally, satellite operating parameters such as navigation data errors, signal availability l anti-spoof failures, and certain types of satellite clock failures are monitored internally within the satellite. If such internal failures are detected, users are notified within six seconds. Other failures detectable only by the control segment may take from fifteen minutes to several hours to rectify.

Both the US Department of Transport and Department of Defense have recognized the requirement for additional GPS integrity for aviation. The development of integrity capabilities to meet flight safety requirements is underway.

Appendix 9

Practical DGPS Systems

1. Short-range Systems

Most GPS receivers, even some hand-helds, now have differential capability built-in, and monitor-quality receivers capable of producing corrections in the correct format are easily obtainable. There are a number of GPS receivers-on-a-card designed to be plugged into one of the expansion slots of a PC that have the ability to act as a DGPS master station. They output formatted DGPS data either to the PC directly or to a dedicated data line and come with the software necessary to do it. At least one such card can be configured through its own software to provide a fully-formatted SC-104 data stream at one of the COM ports of the computer, at any desired speed. It is then a simple matter to arrange for this to modulate a suitable data transmitter, either sitting next to the PC, or via a 'phone line. Some portable calibration systems arrange to have all this on a small platform atop an ordinary surveyor's tripod powered by a car battery sitting underneath, which is quite adequate for a range of 4–5 miles. Of course, there must also be some sort of automatic monitoring system but this could easily be a data-link receiver and a GPS receiver coupled together through a PC that can activate some sort of alarm if the corrected GPS position wanders outside specification. At its crudest, the alarm might simply switch off the data flow to the data link transmitter. An aircraft installation, presuming it already has a suitable DGPS-capable GPS receiver, would require a suitable data link receiver with its aerial, plus an interface to the GPS receiver to be installed.

Expert advice should be sought if it is proposed to set up a local-area DGPS to cover the area around an airfield. In addition to the engineering aspects, there might be questions of safety and legal liability.

It is possible to use the existing marine medium-frequency beacon (transmitting at around 300 kHz) DGPS network in aircraft and one or two successful installations have been made. The existing NDB aerial and receiver should not be used because they are not usually sensitive enough and in any case because a certified installation should not be interfered with in any way.

2. Offshore and Marine Systems

Surveyors working offshore more than a few tens of miles are beyond the reach of shore-based line-of-sight data links. Fortunately this type of work does not always demand the very high reliability required for aircraft landing and data links of lower reliability can be used. Special radio channels in the low, medium and high-frequency bands are used and their occasional unreliability due to thunderstorms and skywaves is accepted. These systems are generally commercial systems available only to those who pay for them, but DGPS is also transmitted over some medium-frequency transmitters in the 300 kHz band that were formerly used only for DF purposes. In all countries except the UK they are run as a national service and are freely available to anyone buying a suitable receiver, but in the UK they are encrypted and run as a commercial service and a decoder must be leased from the operating company.

3. Geostationary Satellite Data Links

Some offshore exploration areas are out of range of both line-of-sight and extended-range MF and HF systems. To cater for this, a number of DGPS systems using geostationary satellite links have been set up. These systems use data links through INMARSAT satellites and transmit data for several monitor sites simultaneously on each link.

One of the best known is the Racal-Decca 'Skyfix' system (Fig A13). This system is controlled through two major control centres at Aberdeen and Singapore, which between them handle no less than sixty monitors scattered over most of the world. The data is sent over a narrow-band data link at a speed of 1,200 bps (2,400 baud signalling rate) and the system achieves a latency of only six seconds in spite of the considerable distances the data has to travel. In order to reduce costs, important in a commercial service, only low-power is transmitted from the satellites because the cost of the INMARSAT link depends on the power used. This requires the receiving vessel to have a standard high-gain INMARSAT-A communications dish aerial on board but since nearly all offshore survey vessels now carry this equipment it does not cause any particular problem, although it precludes its use by aircraft. One way by which aviation might avail itself of this type of service is for an airfield to pick up the INMARSAT down-link and then re-transmit it over a local VHF link for use by aircraft in their locality. This would save each airfield setting up its own system.

Reception of a satellite DGPS data link directly aboard an aircraft needs much higher satellite power because of the impossibility of putting a big high-gain aerial on an aircraft. This makes the system considerably more costly but might be justified if there were a large number of users. Such a system is proposed for WAAS (see Chapter 8). In this case the cost is justified by the very high reliability required.

4. FM Radio Data Links

Many land users need higher accuracy than standard GPS and would like to use DGPS but covering an entire country or continent with a network of special DGPS data links would need a large number of transmitters and be prohibitively expensive. However, most countries already have a dense network of transmitters in the form of radio and television broadcasting transmitters and it is possible to use them for the data link. Apart from almost complete coverage they are also usually high-powered and need only a very simple receiver. All standard European FM transmissions (and many American) now carry digital data via the Radio Data Service (RDS) in addition

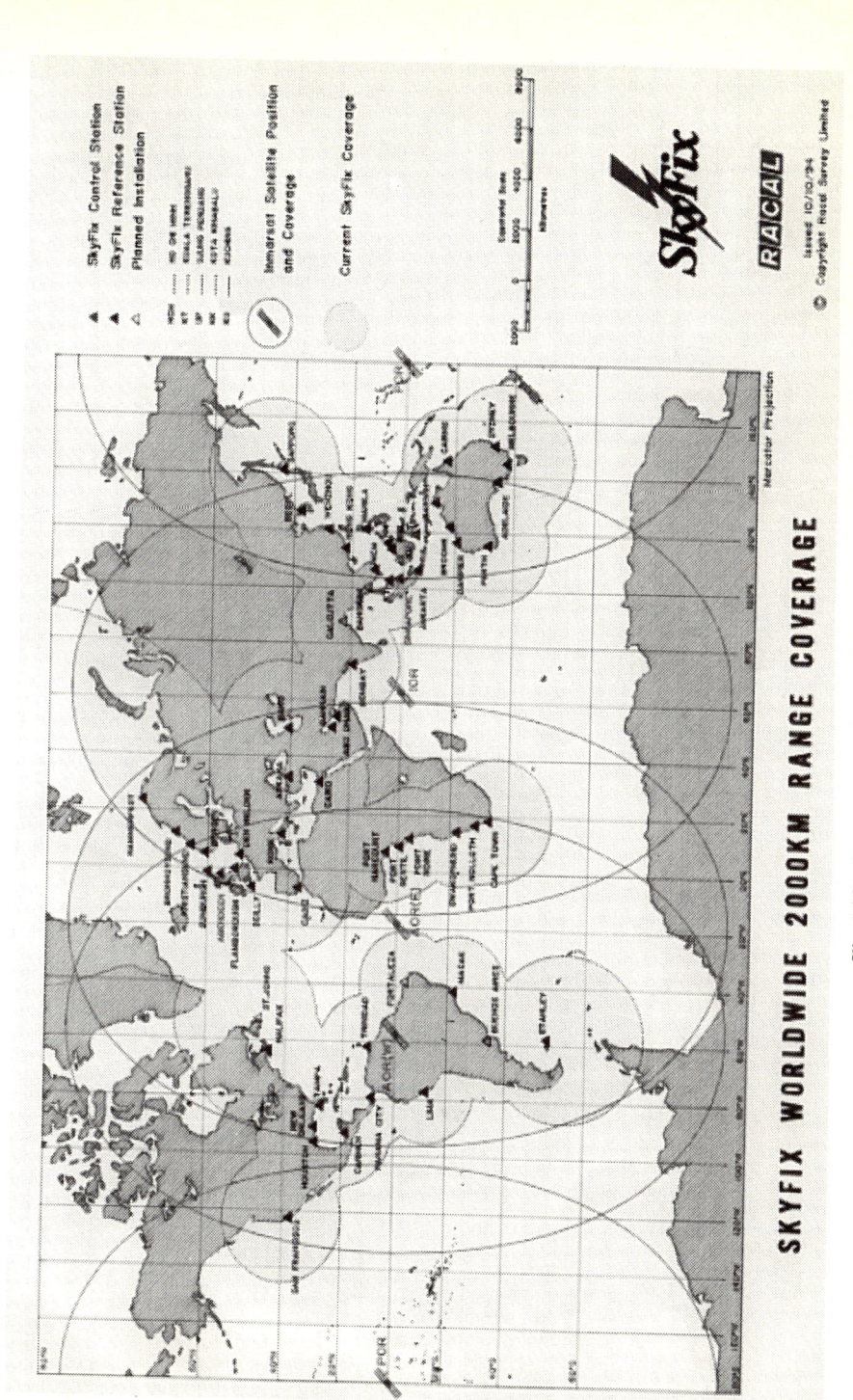

Fig A13 – Racal-Decca 'Skyfix' DGPS System

The Air Pilot's Guide to Satellite Positioning Systems

Applications	Group types which contain this information	Appropriate repetition rate per sec.
Programme identification (PI) code	all	11.4
Programme type (PTY) code	all	11.4
Traffic programme (TP) identification code	all	11.4
Programme service (PS) name	0A, 0B	1
Alternative frequency (AF) code pairs	0A	4
Traffic announcement (TA) code	0A, 0B, 15B	4
Decoder identification (DI) code	0A, 0B, 15B	1
Music/speech (M/S) code	0A, 0B, 15B	4
Radiotext (RT) message	2A, 2B	0.2
Enhanced other networks information (EON)	14A, 14B	up to 2

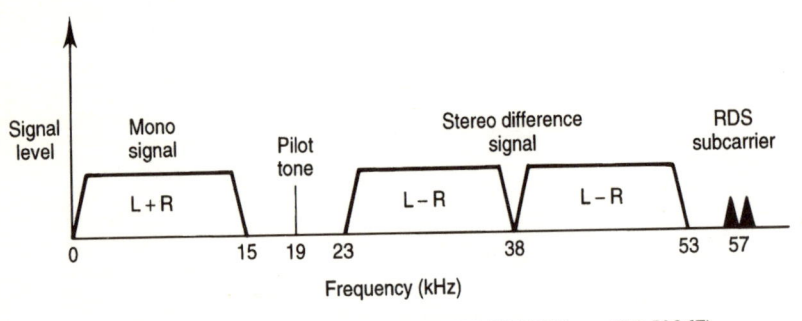

Fig A14 – RDS Message Allocations (CENELEC Spec. EN 50067)

to the normal programme. It is carried on a low-power sub-carrier 57 kHz away from the main programme carrier and does not interfere with it. RDS allows automatic identification and selection of radio services and has additonal information such as traffic and other warnings (Fig A14). Provision is also made for other information which may occupy up to about 25% of the total.

This slightly limited capability for DGPS results in a rather longer latency than can be achieved with dedicated data links. Typical figures are for between ten and twenty seconds when transmitting data for eight satellites. This does not meet the 'all-in-view' requirement but is probably adequate for land vehicles travelling in suburban areas where low-elevation GPS reception is limited. Even to achieve this latency the data signal is usually compressed and makes an assumption that only relatively low-speed vehicles will be using it. If used in an aircraft this might lead to problems not only with latency but also with the corrections themselves being incorrect. Another possible problem for aviators is that the radiation pattern of high-powered FM transmitter aerials is usually strongly concentrated at low elevations towards the horizon. Although the signal might be satisfactory at low altitudes it can fall off much more rapidly than expected at higher altitudes leading to a need to rapidly switch between transmitters. This may or may not be a problem, the difficulty lying not in actually switching radio frequency but in re-synchronising to the new data stream which will have different timing from the previous one.

Appendix 10

List of GLONASS Launches

GLOSNASS Satellite History

No	COSMOS	System	Frequency	Type	Launch date	Operational	End of operation	Orbital Plane	Notes
1	1413			I[3]	12.10.82	10.11.82	12.01.84		
2	1490			I	10.08.83	02.09.83	27.09.84		
3	1491			I	10.08.83	31.08.83	18.10.83		
4	1519			I	29.12.83	07.01.84	27.09.84		
5	1520			I	29.12.83	15.01.83	30.01.86		
6	1554			I	19.05.84	05.06.84	16.08.85		
7	1555			I	19.05.84	09.06.84	25.10.85		
8	1593			I	04.09.84	22.09.84	28.11.85		
9	1594			I	04.09.84	28.09.84	04.09.86		
10	1650			I	18.05.85	06.06.85	29.11.85		
11	1651	1	7	IIa[3]	18.05.85	04.06.85	23.07.87	1	
12	1710	18	4	IIa	25.12.85	17.01.86	16.02.87	3	
13	1711	17	19	IIa	25.12.85	20.01.86	16.05.87	3	
14	1778	2	11	IIa	16.09.86	17.10.86	20.02.87	1	
15	1779	3	20	IIa	16.09.86	17.10.86	15.07.88	1	
16	1780	8	22	IIa	16.09.86	17.10.86	15.06.88	1	
17	1838			IIb[4]	24.04.87	—	—	—	Failed Launch
18	1839			IIb	24.04.87	—	—	—	Failed Launch
19	1840			IIb	24.04.87	—	—	—	Failed Launch
20	1883	6	6	IIb	16.09.87	10.10.87	06.06.89	3	
21	1884	6	6	IIb	16.09.87	09.10.87	20.08.88	3	
22	1885	17	6	IIb	16.09.87	05.10.87	25.01.89	3	
23	1917			IIb	17.02.88	—	—	—	Failed Launch
24	1918			IIb	17.02.88	—	—	—	Failed Launch
25	1919			IIb	17.02.88	—	—	—	Failed Launch
26	1946	7	6	IIb	21.05.88	01.06.88	10.05.90	1	
27	1947	8	6	IIb	21.05.88	04.06.88	19.03.91	1	
28	1948	1	6	IIb	21.05.88	03.06.88	10.06.90	1	
29	1970	17	6	IIc[5]	16.09.88	20.09.88	21.05.90	3	

1	2	3	4	5	6	7	8	9	10
30	1971	18/20[1]	[6]	IIc	16.09.88	28.09.88	30.08.89	3	
31	1972	18/18[2]	[6]	IIc	16.09.88	02.10.88	01.11.91.	3	
32	1987	2	9	IIc	10.01.89	01.02.89	14.03.93	1	
33	1988	3	6	IIc	10.01.89	01.02.89	17.01.92	1	
34	2022	24	[6]	IIc	31.05.89	04.07.89	23.01.90	3	
35	2023	19	[6]	IIc	31.05.89	15.06.89	18.11.89	3	
36	2079	17	21	IIc	19.05.90	20.06.90	—[8]	3	
37	2080	19	3	IIc	19.05.90	17.06.90	—	3	
38	2081	20	15	IIc	19.05.90	16.06.90	18.08.92	3	
39	2109	7	13	IIc	08.12.90	01.01.91	—[9]	1	
40	2110	4	14	IIc	08.12.90	29.12.90	—[9]	1	
41	2111	5	23	IIc	08.12.90	28.12.90	—	1	
42	2139	22	11	IIc	04.04.91	28.04.91	—	3	
43	2140	21	20	IIc	04.04.91	28.04.91	06.01.92	3	
44	2141	24	14	IIc	04.04.91	26.04.91	26.02.92	3	
45	2177	3	22	IIc	30.01.92	24.02.92	12.01.93	1	
46	2178	8	2	IIc	30.01.92	22.02.92	—	1	
47	2179	1	23	IIc	30.01.92	18.02.92	—	1	
48	2204	21	24	IIc	30.07.92	19.08.92	—	3	
49	2205	20	8	IIc	30.07.92	29.08.92	—	3	
50	2206	24	1	IIc	30.07.92	26.08.92	—	3	
51	2234	3	12	IIc	17.02.93	14.03.93	—[9]	1	
52	2235	6	22	IIc	17.02.93	16.03.93	—	1	
53	2236	2	5	IIc	17.02.93	14.03.93	—	1	
54	2275	17[7]	[7]	IIc	11.04.94	—	—	3	
55	2276	17	24	IIc	11.04.94	—	—	3	
56	2277	23	3	IIc	11.04.94	—	—	3	

Notes:
1. 06.08.89. transferred from 18th orbital point to 20th
2. 05.06.89. transferred from 19th orbital point to 18th
3. Satellite lifetime – 1 year
4. Satellite lifetime – 2 years
5. Satellite lifetime – 3 years
6. Data not available
7. Will be moved to the 18th orbital point, with frequency 11
8. Replaced by C-2276
9. Operational but placed in reserve status
10. Individual satellite operating frequencies can be determined by the following formula:

$$f_k = k_0 + \Delta f$$

where $f_0 = 1602$ MHz, $\Delta f = 562.5$ kHz, and k represents the integer between 1 and 24 as shown in the frequency column.

Appendix 11

Comparison of GPS and GLONASS

		GPS	GLONASS
Number of Satellites		24	24
Number of planes		6	3
No of satellites each plane		4	8
Frequencies (MHz)	L1	1565.2–1585.7	1602.5625–1615.5
	L2	1217.4–1237.8	1240–1260
Modulation		Spread-spectrum	Spread-spectrum
Encryption L2		Yes	Yes
Selective Availability		Yes	No
Orbital Period		717.94 min	675.73 min
Semi-major Axis		26,560 km	25,510 km
Inclination		Block 1 63° Block 2 55°	64.8°
Orbital Plane Separ.		120°	120°
Drift per rev.		180°	169.4°
Grd Track Repeat		Block 1=16 orbits Block 2=17 orbits	17 orbits
Spectral Power Density		−41 dBw/Hz	−44 dBw/Hz
C/A Sync. word		8 bits/20 mSec	30 bits/10 mSec
C/A Code Rate		1.023 Mb/s	0.511 Mb/s
C/A code repeat rate		1 mS	1 mS
P code rate		10.23 Mb/s	5.11 Mb/s
P code repeat rate		38 weeks	?
Channel spacing (MHz)	L1	0	0.5625 MHz
	L2	0	0.4375 MHz
Data Rate		50 bps	50 bps
Data Frame length		1500 bits/30 secs	7500 bits/150s
Subframes		5/300 bit	5/1500 bit
Word (lines)		10 per subframe	15 per subframe
Satellite weight		Block 1 = 455 kg Block 2 = 815 kg	700 kg

Appendix 12

The Wide-Area Augmentation System (WAAS)

The Federal Aviation Administration's (FAA) description of WAAS is that it is 'a system consisting of equipment and software to augment the Standard Positioning Service (SPS) of GPS'. WAAS is designed to meet the requirements for a stand-alone navaid for all phases of flight from en route to precision approach. Although WAAS users will primarily be aviators there will probably also be non-aviation users.

WAAS will provide two basic services:
(1) independent integrity data on GPS and the GEO satellites used in WAAS,
(2) a ranging capability additional to, but compatible with, GPS.

1. Method of Operation

GPS satellite data will be received and processed at a number of monitors, referred to as Wide-area Reference Stations (WRSs). These data will be forwarded to data processing sites, called Wide-area Master Stations (WMSs), for processing the integrity, differential corrections, residual errors, and ionospheric information for each monitored satellite. Because geostationary (GEO) satellites are a necessary part of the system, precise orbital information will be generated for them also. This information will be sent to a Ground Earth Station (GES) and uplinked along with the GEO navigation message to GEO satellites. The GEO satellites will downlink this data on the GPS L1 frequency with a modulation similar to that used by GPS. In this way, no additional equipment will be needed by a GPS user to make use of the WAAS system other than a modified software suite in his GPS receiver. This downlink will be differentiated from GPS by using non-GPS, but compatible, PRN codes.

In addition to providing GPS integrity information, WAAS will verify its own integrity and will ensure that it meets its own performance requirements.

The data provided by WAAS will be:

(1) integrity data to ensure that only good GPS satellites are used,
(2) differential correction and ionospheric information data to improve the accuracy of the user's position solution,
(3) ranging data from one or more GEO satellites complementing that already provided by GPS.

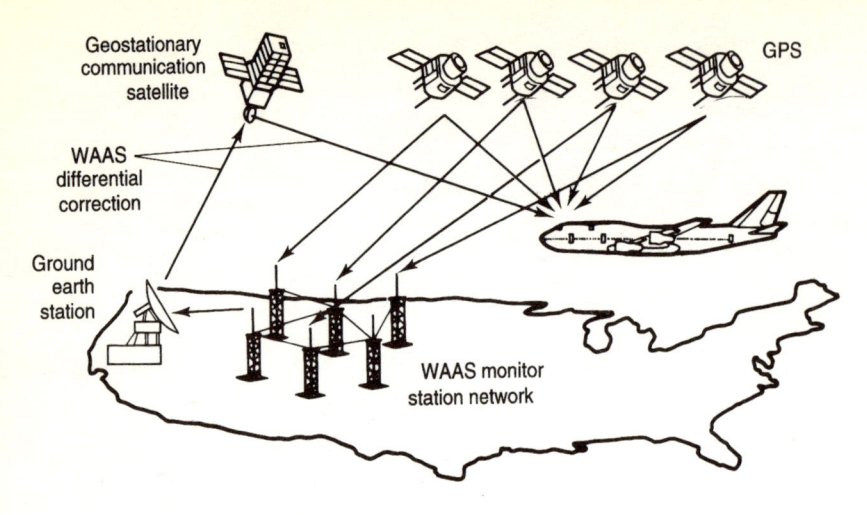

The Air Pilot's Guide to Satellite Positioning Systems

Fig A15 – WAAS Architecture

2. WAAS Objectives

The objectives of WAAS are to provide the necessary:
(1) integrity,
(2) accuracy,
(3) availability, and
(4) continuity of service

to bring GPS up to the level required for a navigation system to be used for all phases of flight including precision approach.

Besides its normal civil mode of operation, WAAS will also have a military emergency state when it will only augment GPS up to the level necessary to meet performance requirements for non-precision approach. This means in practice that while it will continue to provide improved integrity, accuracy will revert back to that provided by standard GPS with SA applied. Like GPS itself, this mode would only be activated under extreme conditions of US national emergency.

3. Specification for WAAS

System Functions

WAAS will perform eight primary functions:
(1) data collection;
(2) the determination of: ionospheric corrections;
(3) satellite orbits;
(4) satellite corrections;
(5) satellite integrity;
(6) the provision of: independent data verification;

Appendix 12

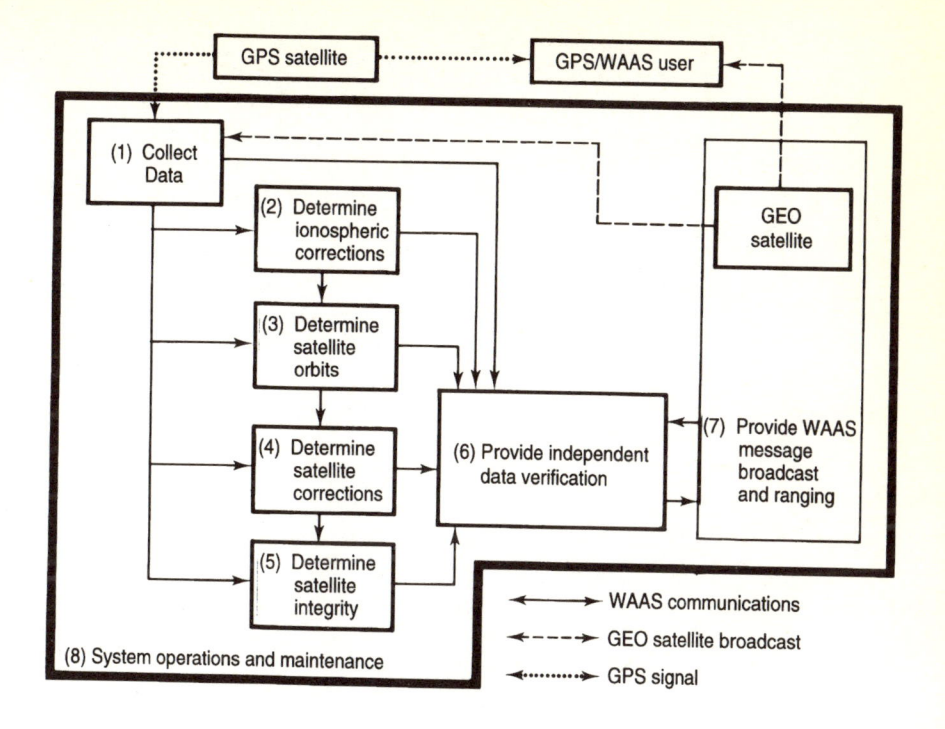

Fig A16 – WAAS Functional Relationships

(7)   WAAS message broadcast;
(8)   GPS-compatible ranging;
(9)   its own system operations.

3.1 Function 1 – Collect Data

Data is received from both GPS and GEO satellites, and atmospheric data obtained to determine tropospheric delay. A check is then performed to verify the reasonability of the collected data. Two independent sets of data are collected: one for Functions 2-5 and one for Function 6.

Inputs:
a. GPS satellite observables;
b. GEO satellite observables;
c. atmospheric observables;
d. equipment location data; and
e. equipment calibration data.

Outputs:
a. GPS L1-C/A pseudo-range measurements (independent sets);
b. GPS L1/L2 code differential measurements (independent sets);
c. GPS satellite navigation data;
d. GEO L1-C/A code pseudo-range data (independent sets);
e. GEO satellite navigation data;

The Air Pilot's Guide to Satellite Positioning Systems

    f. atmospheric data (independent sets);
    g. antenna phase center locations;
    h. receiver L1/L2 differential pseudo-range bias data; and
    i. notification of data that fails the reasonability check.

3.2 Function 2 – Determine Ionospheric Corrections
To determine ionospheric corrections, the GPS L1/L2 code differential measurements just obtained are used. Corrections are produced in the form of ionospheric delays calculated for a number of grid points (IGPs) throughout the coverage area of the system.

    Inputs:
    a. GPS L1/L2 code differential data;
    b. receiver L1/L2 differential pseudo-range bias;
    c. antenna phase centre locations;
    d. GPS satellite navigation data; and
    e. ionospheric grid definition.
    Outputs:
    a. IGP locations;
    b. IGP vertical delay estimates; and
    c. IGP grid ionospheric vertical error (GIVE) data.

3.3 Function 3 – Satellite Orbit Determination
This function determines position, velocity, clock offsets and drifts of all satellites. It also generates the GEO satellite ephemeris and almanac data for the GEO satellite ephemeris message.

    Inputs:
    a. GPS L1-C/A pseudo-range measurements;
    b. GPS L1/L2 code differential measurements;
    c. GPS satellite navigation data;
    d. GEO L1-C/A code pseudo-range data;
    e. GEO satellite navigation data;
    f. atmospheric data;
    g. monitor antenna phase centre locations;
    h. receiver L1/L2 differential pseudo-range bias;
    i. GEO ionospheric data;
    j. GEO satellite planned manoeuvres (manual); and
    k. GPS satellite planned manoeuvres (manual).
    Outputs:
    a. GPS satellite orbit data;
    b. GEO satellite orbit data;
    c. GEO satellite ephemeris message data; and
    d. GEO satellite almanac data.

3.4 Function 4 – Determine Satellite Corrections
From satellite navigation and orbit data from all of the satellites in view including GPS and GEO satellites, precise satellite clock and ephemeris error corrections are determined.

Appendix 12

Inputs:
a. GPS L1-C/A pseudo-range measurements;
b. GPS L1/L2 code differential measurements;
c. GPS satellite navigation data;
d. GEO L1-C/A code pseudo-range data;
e. GEO satellite navigation data;
f. atmospheric data;
g. monitor antenna phase centre locations;
h. receiver L1/L2 differential pseudo-range bias;
i. GPS satellite orbit data; and
j. GEO satellite orbit data.

Outputs:
a. long term corrections;
b. fast error corrections; and
c. User Delay Range Estimates (UDREs).

3.5 Function 5 – Determine Satellite Integrity

This will provide warnings when any satellite or correction should not be used for navigation, or notify the user if there are any satellites or ionospheric grid points that cannot be monitored for any reason.

Inputs:
a. GPS satellite navigation data;
b. GEO satellite navigation data;
c. monitor antenna phase centre locations;
d. satellite long term error corrections;
e. satellite fast error corrections;
f. IGP locations;
g. IGP vertical ionospheric delay estimates; and
h. IGP GIVE data.

Outputs:
a. data to generate 'Don't Use' messages;
b. data to generate 'Not Monitored' messages; and
c. list of satellites that should be in view but whose signals are not being received.

3.6 Function 6 – Independent Data Verification

This function verifies the integrity of all data prior to transmission and validates that data while it is in use. 'Verification of data' is defined to be a process whereby data is either:

(1) compared to data derived from independently observed measurements or,
(2) combined with independently observed measurements and compared to the expected result.

Inputs:
a. GPS L1-C/A pseudo-range measurements;
b. GPS L1/L2 code differential measurements;
c. GEO L1-C/A code pseudo-range data; and
d. atmospheric data.

e. GEO satellite navigation data;
f. GPS satellite navigation data;
g. monitor antenna phase centre locations;
h. receiver L1/L2 differential pseudorange bias;
i. IGP locations;
j. IGP vertical delay estimates;
k. IGP GIVE data;
l. GPS satellite orbit data;
m. GEO satellite orbit data;
n. GEO satellite ephemeris message data;
o. GEO satellite almanac data;
p. satellite long-term corrections;
q. satellite fast error corrections;
r. satellite UDREs;
s. data for 'Don't Use' messages; and
t. data for 'Not Monitored' messages.

Outputs:
a. indications for 'Don't Use' and 'Not Monitored' satellites or IGPs;
b. Satellite fast corrections;
c. Satellite long-term clock and ephemeris corrections;
d. IGP locations;
e. IGP vertical delay estimates;
f. IGP GIVE data;
g. GEO satellite ephemeris data;
h. GEO satellite almanac data; and
i. Satellite UDREs.

3.7 Function 7 – WAAS Message Broadcast and Ranging

This assembles a GPS-like ranging source and integrity and correction signal and sends it to a GEO satellite for re-broadcast.

Inputs:
a. indications for 'Don't Use' and 'Not Monitored' satellites or IGPs;
b. satellite fast corrections;
c. satellite long-term clock and ephemeris corrections;
d. IGP vertical delay estimates;
e. IGP GIVE data;
f. GEO satellite ephemeris data;
g. GEO satellite almanac data,
h. satellite UDREs;
i. IGP locations;
j. list of GPS/WAAS satellites;
k. UTC/WAAS Network Time (WNT) offset; and
l. GEO satellite testing mode command.

WAAS Message Format
Basically a 500 symbols-per-second data stream with all alarm messages repeated within one second. This is added to a unique 1.023 MHz 1023 bit Gold Code assigned

Appendix 12

to each GEO satellite, thus producing a C/A code-like signal. This is uplinked to the GEO satellite on a suitable non-GPS related radio-frequency and the satellite then translates it to the WAAS broadcast carrier frequency of 1575.42 MHz (GPS L1) and re-transmits it.

Outputs:
a. WAAS messages;
b. ranging signals; and
c. WAAS signal quality parameters.

3.8 Function 8 – System Operations and Maintenance

The control, monitor, and maintainence function of WAAS itself. It is comprised of four subfunctions:

(1) system operations and maintenance data collection;
(2) system monitor and control;
(3) corrective maintenance; and
(4) preventive maintenance.

3.9 The WAAS Data Message

This allows for as many as 210 satellites, of mixed orbits, and full ionospheric refraction correction.

Table A5: Message Types

Type	Contents
0	Don't Use this GEO for anything (for WAAS testing)
1	PRN Mask assignments, set up to 52 of 210 bits
2	Fast corrections
3–8	Reserved for future messages
9	GEO Ephemeris Message (X, Y, Z, time, etc.)
10	Reserved for future messages
11	Reserved for future messages
12	WAAS Network/UTC offset parameters
13–16	Reserved for future messages
17	GEO Satellite Almanacs
18	Ionospheric Grid Point Mask # 1
19	Ionospheric Grid Point Mask # 2
20	Ionospheric Grid Point Mask # 3
21	Ionospheric Grid Point Mask # 4
22	Ionospheric Grid Point Mask # 5
23	Reserved for future messages
24	Mixed Fast corrections/Long-term satellite error corrections
25	Long-term satellite error corrections
26	Ionospheric error corrections
27–63	Reserved for future messages

The Air Pilot's Guide to Satellite Positioning Systems

The Type 1 message indicates for which satellites corrections are contained in the broadcast. It has to take account of every navigation satellite because its use of multiple monitors over a very wide area means that the total system could conceivably have all twenty-four GPS satellites 'in view' simultaneously, and there may be additional satellites in the system such as twenty-four GLONASS, geostationary, and others. In all it is designed to allow for up to 210 PRN assignments.

The actual corrections for those satellites is contained in the subsequent messages 2, 24 and 25, divided into fast short-term corrections and longer-term, slower, corrections.

System data for navigation payloads aboard geostationary satellites is sent via messages 9 and 17.

One of the major differences from the RTCM SC-104 system is that WAAS attempts a much more comprehensive ionospheric error correction. The vertical ionospheric delay at 929 ionospheric grid-points within the coverage of each WAAS broadcast are calculated from observed data and data for blocks of 186 are transmitted. The user receiver can obtain an up-to-date picture of ionospheric behaviour in the local area and can correct for it.

There is considerably more data to be transmitted in this system than in RTCM and therefore the data rate is higher at 250 bps. This is not the actual signalling rate – because of digital error checking (Forward Error Correction – FEC) it becomes 500 bps modulation.

Appendix 13

The European Geostationary Navigation Overlay Service – EGNOS.

If the benefits of the WAAS system just described are to be obtainable world-wide then WAAS-like systems must be installed outside the USA and be fully compatible with it. Also, if the promised economic benefits are to be achieved then they must be made simultaneously available as widely as possible. Airlines do not like having to fit different sets of equipment for different areas of the world! EUROCONTROL (see Appendix 3) and the US Federal Aviation Authority are therefore collaborating on the design of their GPS augmentation systems in order to make them technically compatible and inter-operable. The aim is that although different authorities may operate these systems and they may have different names there will be no difference as far as the user is concerned.

In Europe, a Tripartite Group of Eurocontrol, the European Space Agency, and the European Commission has been formed with the task of designing and implementing the European version of WAAS, EGNOS. Eurocontrol will take responsibility for operational aspects, ESA engineering, and EC finance. It is expected that although EGNOS will be its first project, the Group will also in due course consider the development of full satnav systems.

1. Technical aspects of EGNOS

In late 1995 and 1996, Inmarsat will commence launching its Generation III geostationary communications satellites, each of which will carry a navigation transponder. Three of these will be able provide service to Europe – two on the Atlantic (East and West stations) and one on the Indian Ocean station. To maintain integrity of the overlay service itself augmentation signals must be receivable from at least two GEO's simultaneously for each area, and for Europe these will be initially the Atlantic East and Indian Ocean satellites, with some service from the Atlantic West satellite (Fig A17). For the USA, the FAA will be using the Atlantic West and Pacific Ocean satellites and thus there is overlap in that the Atlantic West satellite serves both the USA and Europe. For this reason alone signal format and data must be identical in both areas even if it is uploaded by different controlling authorities. There will also be overlap for Far Eastern services, which will be obtained from the Indian and Pacific Ocean satellites plus a new satellite halfway between them, but the controlling authority for this service has not yet been determined.

The general technical outline of how the EGNOS system will work is very similar

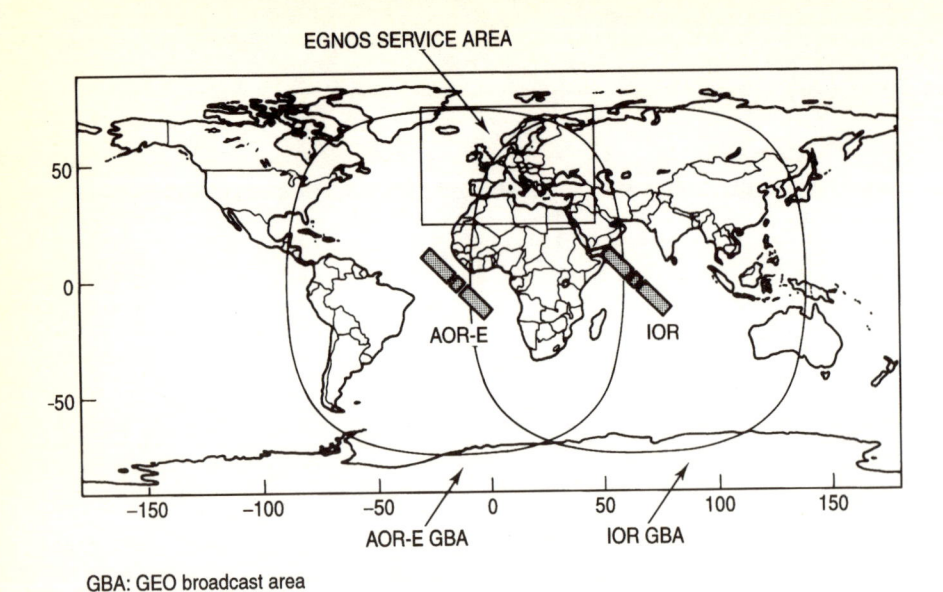

The Air Pilot's Guide to Satellite Positioning Systems

Fig A17

to those already given in the Inmarsat and WAAS sections. In particular, the data and data format will be identical to WAAS. The primary area will of course be Europe but there will be considerable overlap outside its borders. As long as the basic monitoring station data is valid the signal will be usable but since in the first instance all the monitors will be in, or near, Europe service will not necessarily be maintained (for instance) over Africa although the GEO satellite signal will be receivable there. The first four monitors, designed mainly for integrity improvement purposes, will be located in the general areas of Iceland, Northern Scandinavia, the Eastern Mediterranean, and the Canary Is. and there are plans for additional monitors outside Europe later. More monitors located geographically within the basic four will enable better accuracy enhancement.

For a user located at 60°N and 10°E, the single additional ranging signal from the Atlantic East satellite is predicted to increase GPS PDOP from 2.55 (90%) to 1.98 (90%). This may seem a small improvement but in fact it eliminates one or two periods of much worse PDOP. As regards accuracy improvement, while the major effect is to remove the effects of GPS S/A the basic GPS SPS accuracy is also increased and early tests have shown that a 95% accuracy of 5.4 metres horizontally and 12.6 metres vertically should be obtainable. The upload station will be that of France Telecom at Aussaguel, Brittany, with a standby provided by the Deutsche Telecom station at Raisting.

One of the major differences between WAAS and EGNOS is that EGNOS has a primary service area much further North than WAAS and while this does not cause any difficulty with basic GPS enhancement, it raises the question of whether GEO satellite signals are satisfactorily receivable aboard aircraft far to the North. Some

Appendix 13

testing has already been done on this, resulting in the discovery that in fact they can be quite well received as far North as 84°, where the theoretical elevation of a GEO satellite is below the horizon.

2. Development Schedule

A considerable amount of work is already (1995) in progress, with Eurocontrol and the FAA working closely together. There will be three steps to Phase 1 – Initial Operational Capability:

> Step 1 – The ranging facility and associated ground organisation will be put into place.
> Step 2 – A Ground Integrity Channel (GIC) and Integrity Monitors will be added.
> Step 3 – Wide-Area Differential Corrections and appropriate monitors will complete the system.

The time scale for these steps to be completed is from 1995 to 1999.

Phase 2 will be Full Operational Capability. With the completion of Phase 1 in 1999, there will be a need for considerable testing and possibly modification to ensure reliable operation, scheduled to take place between 1999 and 2002. Provided this is satisfactory, Phase 2 will be completed by 2002 and EGNOS will be declared fully operational by 2002.

Appendix 14

Institutions Involved in the Operation of Aviation Navaids

1. International Civil Aviation Organisation (ICAO)

The International Civil Aviation Organisation is a specialised agency of the United Nations set up by the Chicago Convention of December 1944, and changed from Provisional to full status in April 1947. It has 160 members. Its aims are to develop the principles and techniques of international air navigation and to foster the planning and development of international air transport.

Its charter includes the encouragement of the development of air navigation facilities, promotion of safety of flight, and the standardisation of communications systems and air radio-navigation aids. It has sponsored a number conventions dealing with international air law, and regularly has international meetings for the purpose of standardising radio navaids, amongst others.

It has gone considerably further than any other organisation towards the establishment of common radio navaid standards.

Annex 10 to the Convention specifies in detail the recommended standard radio navaids and procedures for using them.

It thus has in being a complete mechanism for specifying and setting standards for the international use of aviation radio navaids.

1.1 ICAO FANS Committee

In 1983, ICAO set up the Future Air Navigation System Committee (FANS) to lay the foundations for 'the development of air navigation for international civil aviation over a period of twenty-five years'. Four years later the FANS Committee concluded that 'the exploitation of satellite technology to provide Communications, Navigation and Surveillance (CNS) services to civil aviation on a global basis is the only viable solution that will enable the shortcomings of the present air navigation system to be overcome and fulfil the needs and requirements of the foreseeable future'. The proposed FANS system embraced the satellite-based CNS concept and greatly improved arrangements on the ground for the purpose of Air Traffic Management (ATM).

After consideration of the final report of the FANS Committee, the ICAO Council established in 1989 the 'Special Committee for Monitoring and Co-ordination of Development and Transition Planning for the Future Air Navigation System' which is known as 'FANS II'. Its terms of reference include the requirement to 'Identify and make recommendations for acceptable institutional arrangements, including funding, ownership and management issues for the global future air navigation system'.

In September 1991, the global FANS concept was endorsed by the 10th Air

Appendix 14

Navigation Conference (ANC) of ICAO. The Conference also agreed general Guiding Principles on institutional and legal aspects of FANS, and particular Guidelines on the communications part (Aeronautical Mobile Satellite services) of the FANS concept.

At the same time, the United States and the USSR offered to provide GPS and GLONASS to civil users free of charge for the next ten and fifteen years respectively. Since this offer covered only a limited period, the interest in a civil system for the long-term continuity was recognised. Therefore the 10th ANC recommended in relation to the Global Navigation Satellite System (GNSS), the satellite navigation part of the FANS concept: 'That ICAO, as a matter of urgency, develop the Institutional arrangements (including Integrity aspects) as a basis for the continued availability of GNSS for civil aviation' (Recommendation 4/4).

The Legal Committee of ICAO has been requested to consider the establishment of a legal framework, particularly in view of the long term ICAO GNSS strategy. The 28th Session of the Legal Committee of ICAO met in May 1992 and considered the guiding principles agreed by the 10th ANC. Specific guidelines for acceptable institutional arrangements relating to the implementation of GNSS, proposed in April 1992 by FANS, were approved and added to the guidelines on AMSS.

In a clarification on the task of the Legal Committee in respect of the legal framework for GNSS, which was recently presented by the Secretary General of ICAO, some backgrounds and elements of the legal framework are provided for. These considerations reflect many of the issues dealt with in ICAO documents on FANS and point to basic requirements of the civil aviation community in respect of Institutional arrangements for GNSS.

> 'The evolution of GNSS into a sole-means navigation system for civil aviation that is internationally acceptable generates a number of institutional and legal questions which need to be resolved in the near future in order to allow GNSS to develop to its full potential. GNSS can meet aviation navigation and surveillance needs on a full-time, worldwide basis, provided that its services are available in a way that is internationally acceptable and integrates the various space, air and ground components into a cohesive system which is user-transparent. It must achieve an equitable mechanism for recovering costs and be held accountable for the required level of service.'

(General Work Programme of the Legal Committee, Council 138th session, C-WP9725, 18 Feb, 1993).

> 'The main institutional element in the global introduction of sole-means GNSS is related to the provision of assurances to all users of the reliable quality of information and of assurances to sovereign States that the service will be continuous.'

One of the issues to be addressed by the Legal Committee in this respect is:

> 'The definition of internationally acceptable institutional arrangements that are deemed necessary for the provision of a long-term GNSS system which is designed to meet civil aviation needs, taking into account the guidelines for acceptable institutional arrangements relative to the implementation of GNSS as approved by the 28th session of the Legal Committee. The arrangements

should include provisions relating to ownership, operation and control of GNSS components, system funding, costs and cost-recovery and liabilities.'

The 'Guidelines for acceptable institutional arrangements relative to the implementation of Aeronautical Mobile-Satellite Services (AMSS) and Global Navigation Satellite System (GNSS) for Civil Aviation' as agreed by the FANS II Committee in April, 1992 provide a practical basis for defining specific issues to be taken into account to derive acceptable institutional arrangements. Taking these Guidelines and other relevant ICAO (Legal Committee) papers into account it should be possible to provide an institutional model which suits the long-term requirements of the civil aviation community.

(Note: The above has been abstracted from WP 77, FANS II Meeting, London 10–18 May 1993.)

1.2. All Weather Operations Panel (AWOP)

This Panel covers the applications of radio navaids used for precision approach and landing. In addition to its use for en-route navigation, GNSS has been proposed as a system that with some augmentation might be usable as a partial or complete replacement for ILS/MLS.

ICAO Annex 10 standards (Chapter 2) provide for both ILS and MLS as standard non-visual aids to final approach and landing and establish 1 January 1998, as the ILS end-of-protection date when MLS will become the primary system. Any prolongation of the use of the ILS after this will be discouraged. ILS service is to cease as an international standard by 1 January 2000. There is no mention of GNSS in this Annex, but some States view it as an alternative and have slowed down acquisition of MLS although it should be stressed that MLS is the approved future precision landing system.

If GNSS proves to have applications as a landing system, the provisions of this Annex will need reconsideration and it is likely a decision on this will be taken in 1995.

At its meeting in January 1993 AWOP was requested to assess and make appropriate recommendations as to:

a. the application of new technology systems with regard to approach and landing operations, and study in particular, the possible extent of the use of global navigation satellite system (GNSS) for approach and landing operations;
b. the operational and technical possibility of extending the required navigation performance (RNP) concept to include approach and landing operations.
c. ICAO's tentative programme of meetings for 1995 includes a special meeting at the global level which will cover ILS/MLS transition. AWOP was requested to provide material to support the convening of this meeting.

Appendix 14

2. ECAC and Eurocontrol

2.1. European ATC Harmonisation and Integration Programme (EATCHIP)

The European Civil Aviation Conference (ECAC), which has thirty-two member states, has adopted a strategy for the 1990s known as EATCHIP. Eurocontrol was given the responsibility of implementing it and were instructed to co-operate with the CEC and non-Eurocontrol ECAC states in so doing.

One reason for this was the fact that there are in Europe fifty-one Air Traffic Control Centres, using twenty-two different computer operating systems. Some of the air traffic delays are due to the inability of one system to 'talk' to another one. Harmonisation into one system is an aim of this programme and another is the implementation of new route and airspace structures and the development of common procedures and system support functions.

The adopted strategy was:
Phase 1   Appraisal and Evaluation. Completed 1991.
Phase 2   Programme Development. Completion mid-1993.
Phase 3   Acquisition and Implementation. Completion 1995/1998.
Phase 4   Implementation of the future air traffic management system, 1995–2000+.
Operational Objectives are:
a. Route network and airspace structure to be optimised from 1993.
b. Widespread application of RNAV from 1993.
c. Comprehensive radar coverage throughout ECAC by 1995.
d. 5 nm en-route separation in high-density areas by 1995.
e. ATC system harmonisation in high-density
   areas by 1995. In all areas by 1998.
f. Automatic data comms between ATC centres by 1998.
g. Mode S operational 1998 onwards.

3. The International Maritime Satellite Organisation (INMARSAT)

INMARSAT operates the space segment of a world-wide system of satellite communications for predominantly maritime purposes although aviation is now also included. It is an international treaty organisation open to all states and those sixty-seven states who have signed its Convention all have a say in its policies and direction. Its Convention already covers the provision of satellite navigation services. In accordance with this provision it has already carried out experiments in transmitting navigation signals over one of its satellites, and is actively participating in GPS integrity studies.

In 1991 it decided to incorporate into four of its next generation satellites a payload designed specifically for navigation applications. These satellites will be launched in 1995. It has requested the aviation community to provide it with guidance as to international requirements in respect of integrity and other augmentations.

4. The European Organisation for Civil Aviation Equipment (EUROCAE)

This organisation exists to define and recommend standards for aviation equipment and is in many ways a parallel to the US's RTCA organisation. It has many Working Groups on various equipments but the only one concerned specifically with satellite navigation is its Working Group 28 which has been in existence for some years charged with the development of Minimum Operational Performance Standards for GPS, working in conjunction with RTCA SC-159. That work was completed in 1992 and WG-28 has now been re-convened to consider other aspects of GNSS for which it has formed six sub-groups.

They are:

Sub-group	RTCA Equiv.	Subject
1	WG-5	GNSS Integrity
2	WG-2	RGIC
3	?	Not defined
4	WG-4	Precision Landing and surface surveillance
5	WG-1	RAIM
6	WG-3	GLONASS/Other

References

General

For detailed descriptions of all aspects of GPS the three special volumes of *Navigation*, the journal of the US Institute of Navigation, published in 1980–1982, should be consulted.

The engineering specification for GPS is contained in ICD-GPS-200 published by ARINC Research Corporation, USA.

The signal specification for the GPS Standard Positioning Service is given in 'GPS Standard Positioning Service Signal Specification' published by the US Department of Defense. (Nov. 1993).

General GPS system information is given in: 'Technical Characteristics of the Navstar GPS' (June 1991) based on NATO document STANAG 4294 and published by the NATO GPS Technical Support Group.

GPS user equipment information can be found in: 'Navstar GPS User Equipment' (February 1991) (also known as ANP-2) compiled by the NATO GPS team at the USAF GPS Joint Program Office.

The engineering specification for GLONASS is given in: 'Global Satellite Navigation System GLONASS Interface Control Document' published by ISDE/Glavkosmos/RPAM Moscow, Russia.

Official documents:
'Federal Radionavigation Plan', 1992. US Department of Transportation, Washington, DC.

US 'Specification for the Wide-Area Augmentation System', 1994. US Federal Aviation Authority, Washington, DC.

Copies of all the above are available for consultation by members of the Royal Institute of Navigation in the RIN library.

Up-to-date information on all satnav systems is continuously available to members of the U. K. Civil Satnav Group via its Bulletin Board, Newsletter and Fax service.

Address: UKCSG, c/o The Royal Institute of Navigation, 1, Kensington Gore, London, SW7 2AT.

Detailed Information

The following Conferences specialised in satellite navigation:

USION Satellite Division International Technical Meetings:
1988 Colorado Springs.
1989 Colorado Springs.
1990 Colorado Springs.
1991 Albuquerque.
1992 Albuquerque.
1993 Salt Lake City.
1994 Salt Lake City.

Royal Institute of Navigation Conferences:
1989 NAV 89 London.
1991 NAV 91 London.
1994 DSNS 94 London.

German Institute of Navigation:
1991 DSNS 91 Braunschweig.

Netherlands Institute of Navigation:
1993 DSNS 93 Amsterdam.

'Proceedings' of all the above are in the RIN Library and together contain some 1,200 papers on all aspects of satellite navigation.

Historically important papers:
1. 'An Advanced Space Surveillance System', NRL Memo Report, 8 Feb. 1961.
2. T. B. McCaskill, J. A. Buisson, and D. W. Lynch. 'Principles and Techniques of Satellite Navigation Using the Timation II Satellite.' NRL Report 7252, 17 June, 1971.
3. P. S. Jorgensen, 'Navstar/GPS 18-Satellite Constellations'. US I.O.N. Annual Meeting, Monterey (Cal.) June 1980.
4. A. J. Van Dierendonck et al. 'The GPS Navigation Message' *Navigation*, Vol 25, No. 2, 1978.
5. J. Beser and B. W. Parkinson. 'The Application of Navstar differential GPS in the Civilian Community,' *Navigation*, Vol. 29, No. 2. 1982.

Papers not contained in the documents listed above:
1. 'Geostationary Overlay – A Preliminary Cost Estimate'. INMARSAT, April 1989.
2. Nagle, J.R. and Kinal G. 'Geostationary Repeaters' IEEE PLANS 90 Conference, Las Vegas, March 1990.
3. 'European Complement to GPS'. Centre National d'Etudes Spatiales, Feb. 1990.
4. Durand, J.M. 'European Extension to GPS (CE-GPS)and NAVSAT'. Location and Navigation Satellite Systems Conference, Tolosa, 1989.
5. O'Neill, G.K. 'Geostar – Triad Updated'. '*Pilot*' magazine, October 1984.
6. Omnitracs brochure, Qualcomm Inc. 1990.

GPS
1. James, L. 'GPS Satellite Production Status'. US Institute of Navigation meeting, January, 1986.

GLONASS
1. Russian Institute of Radionavigation and Time (RIRT) submission to the EC Study on Satnav for Europe, 1994.
   Dr. Pyotr P. Bogdanov,
   Dr. Sergey S. Pospelov,
   Dr. Irina G. Pushkina,
   Dr. Vladimir V. Korniyenko.
2. Russian GLONASS Presentation at the RTCA GNSS Task Force Meeting, August 12 and 13, 1992. RTCA Digest No. 93.
3. Russian Astronautics Changes an Orbit and Face Earth Problems. S. Leskov. *Izvestiya*, No.28, 13 February, 1993, p. 15 (in Russian).
4. ISDE information provided with the assistance of Ashtech, Inc. *GPS World*, Febr. 1991, p. 39.
5. 'Current Status and Prospects for GLONASS Frequency/Time Support.' B.N. Balyasnikov, P.P. Bogdanov, A.G. Gevorkyan, Y.G. Gouzhva. 'Problems of Radioelectronics', Ser. general problems, 1990, issue 8 (in Russian).
6. 'High-precision Time and Frequency Dissemination with GLONASS.' Y.G. Gouzhva, A.G. Gevorkyan, A.B. Bassevich, P.P. Bogdanov. *GPS World*, July 1992, p. 40-49.
7. Joint US/USSR Satellite Navigation Studies. R.Hartman. *GPS World*. Feb.1992, p. 26–36.

Differential
1. 'Time Division Multiple Access Differential GPS.' Van Dierendonck, A.J. R.T.C.M. paper 22.
2. 'Application of DGPS to Marine Vessel Dynamic Positioning'. Denaro, P. Yoerger, D. Tau Corp., Los Gatos, USA.

Civil Systems and GNSS
1. 'Satellite-based Navigation Technology Update' Nagle, Dierendonck, Kinal. PLANS 94.
2. 'Omnitracs – Technical Definition for the EUTELTRACS System'. Omnitracs Corp. Inc. 1993.
3. *Iridium* – Motorola publicity information, 1994.

Index

1
10th Air Navigation Conference (ANC) of ICAO, 187

A
absolute time, 34, 38
active transponder, 130
aerial, 77
Aeronautical Mobile-Satellite Service (AMSS), 188
Air Traffic Management (ATM), 186
airfield approach pseudo-lite, 48
Airy Spheroid, 149
All Weather Operations Panel (AWOP), 188
almanac, 68
altitude, 15
ambiguity, 134
amplifier, 77
angle of arrival, 50
apogee, 30, 44
area of uncertainty, 18
Argo, 94
artificial grid, 16
ascending node, 140
astro-geodetic measurement, 147
atmospheric refraction, 45, 52
atomic clock, 38
augmentation, 118
    satellite, 31
autonomous capability, 63
autonomous operation, 42
availability, 118
aviation MF beacons, 92

B
Baikonur Space Centre 102

bearing, 15
bit parity error, 47
Block I (GPS satellite), 60
Block II, 62
Block IIR, 42, 63
British Telecom, 107
broadcast ephemeris, 46, 70

C
caesium clock, 60
Cape Canaveral, 71
carrier phase tracking, 81
Cartesian system, 137
cellular network, 114
central control station, 88
chipping rate, 162
circle of position, 32
clock calibration, 21
    drift, 42
coarse/acquisition (CA), 67
code-tracking, 82
Communications Navigation and Surveillance (CNS), 186
COMSAT, 107
convolutional encoding, 92
    interleaving, 92
corner reflector, 130

D
data link, 89
    bit error rate, 92
data output, 83
    protocol, 83
    transmission to user, 48
datum, 15, 148
Decca, 14, 38, 43, 135
Defense Mapping Agency, 70

Index

Defense Navigation Satellite System, 58
Delta II launcher, 62
density of ionisation, 49
Differential GPS, (DGPS), 75, 85
   data-link, 82
   error source, 95
   input, 82
difference of range, 133
differential time of arrival, 35
digital correction, 39
   height map, 111
Dilution of Precision (DOP), 43
direct ranging, 39
directional measurement, 22
Discos, 28
dish aerial, 36
DME, 22
Doppler/Fizeau principle, 143
Doppler-shift, 28, 32
drift of perigee, 65

E
Earth's sphericity, 55
Earth rotation, 96
electronic map, 84
ellipsoid, 147
ellipticity, 138
Encryption, 67
ephemeris data, 45
European Civil Aviation Conference (ECAC), 189
European Datum of 1950 (ED50), 148
European Geostationary Navigation Overlay Service (EGNOS), 183
European Organisation for Civil Aviation Equipment (EUROCAE), 190
European Terrestrial Reference Frame 1989 (ETRF89), 151
Eutelsat, 112
Euteltracs, 48, 111
external time-base, 38

F
fast switching, 80
Fessenden, 13
first point of Aries, 140
fix accuracy, 18

fix behaviour, 20
FM transmission, 168
frequency, 132
   allocation, 157
Future Air Navigation System Committee (FANS), 186

G
Geocentric Cartesian System, 27
   height, 32
geodetic transformation error, 54
Geographic Information System, 152
geographical correction, 87
geoid, 16
geoidal surface, 32
Geoloc, 94
Geometric Dilution of Precision (GDOP), 43
geometric effect, 21
Geostar, 32, 110
Geostationary Earth Orbit (GEO), 26, 30
Global Navigation Satellite System (GLONASS), 14, 48, 99, 117
GLONASS-M, 106
GPS Integrity, 75
gravitational attraction, 26
   field, 47
Greenwich, 38
ground control system, 69
Ground Earth Station (GES), 175
gyroscope, 25

H
half-rate coding, 92
hand-over word (HOW), 67, 162
harmonics, 78
height, 17
Hertz, 13
HF radio, 93
Highly Elliptical Orbits (HEO), 30
Hohmann transfer orbit, 44
Horizontal Dilution of Precision (HDOP), 43
horn aerial, 112
hydrogen maser clock, 60
hyperbolic, 133
hyperboloids, 39
Hyperfix, 94

Index

I
ideal geometry, 43
ILS, 124
inclination, 138
inertial system, 38
Inmarsat, 78, 107
   augmentation system, 60
integrity, 118
   Beacon Landing System (IBLS), 126
   checking, 86
Intermediate Circular Orbit (ICO), 29
internation time-base, 38
International Civil Aviation Organisation (ICAO), 186
international scientific designator, 60
International Telecommunications Union (ITU), 155
inter-satellite link, 39
   timing system, 39
   tracking, 42
integration with ground-based system, 48
ionised layer, 49
ionospheric refraction, 45
   grid-points, 182
   reflection, 49
issue of data (IoD), 90

K
Kalman filter, 70
K-band, 111
Keplerian System, 26, 137

L
land earth station (LES), 107
land-based system, 17
lane identification signal, 82
laser retro-reflector, 105
latency, 91
latitude, 16
Locstar, 32, 46, 110
longitude, 16
longitude of the ascending node (LAN), 69, 140
Loran, 14, 38, 43, 134
Low Earth Orbit (LEO), 27
lunar effect, 45

M
magnetic storm, 94
Marconi, 13
marine radiobeacons, 92
mass-centre of the Earth, 34
master, 133
Master Control Station (MCS), 48, 69
microwave landing system (MLS), 117
MLS, 94, 124
monitor, 87
   clock bias, 96
Morse code, 159
Mottez, 14
multiflex, 80
multipath, 77
multiple satellite system, 35

N
National Grid, 148
National Marine Electronic Association (NMEA), 83
navigation payloads, 128
Navigation Technology Satellite One (NTS-1), 58
NTS-2, 58
NTS-3, 58
Notice Advisory to Navstar Users (NANU), 72

O
Omega, 134
omni-directional aerial, 37
Omnitracs, 32, 111
one-way ranging, 46, 131
Operational Control System (OCS), 69
orbit, 25
Ordnance Survey, 57
Ordnance Survey (Scientific Network) 1980 (OS[SN]80), 150
Ordnance Survey Great Britain 1936 (OSGB36), 149

P
packet network, 112
parallel receiver, 80
   tracking, 80
PDME, 22
perigee, 139

Index

permitted user, 157
phase, 132
　measurement, 81
polar wander, 45
pole, 16
polyhedron, 43
Position Dilution of Precision (PDOP), 43
position line, 17
positioning system, 23
PR number (PRN), 60, 65
precise ephemeris, 46, 70
Precise Positioning Service (PPS), 164
precision approach, 117
pressure altitude, 111
Primary Triangulation of Great Britain, 149
primary user, 156
processing algorithms, 96
Project 621B, 58
propagation – Multipath, 53
　refractive effect, 48
　loss, 68
Proton booster, 100
pseudo-Keplerian, 68
　random noise modulation (PRN), 59
　random noise spread-spectrum (PRNSS), 65
　range correction, 88
　ranging, 39
　ranging two-dimensional system, 40
　satellite, 61
Pulse/8, 94
　code modulation (PCM), 160

Q
quadrifilar helix, 77
quantum optical station (QOS), 102

R
Racal-Decca Skyfix, 168
Radiation, 28
radio altimeter, 111
　astronomy, 102
　Date Service (RDS), 168
　direction-finding, 156
　Regulations 2020, 155

radiodetermination, 156
　satellite service, 156
　service, 156
radiolocation, 156
radionavigation, 156
　satellite service, 156
random error, 20
range, 15
　measurement, 22
　measurement using carrier phase, 131
receiver aerial, 76
　Autonomous Integrity Monitoring (RAIM), 19, 122
　clock drift, 73
　noise, 73
reflection efficiency, 130
relativistic effect, 47
　time shift, 45
Right Ascension of the Ascending Node (RAAN), 140
round-trip time delay, 37
rubidium atomic standard, 42

S
satellite, 16
　aerial, 76
　clock drift, 73
　data link, 93, 127
　ephemeris, 48
　management data, 47
　motion, 22
　positioning, 17
　power, 68
S-band, 111
scintillation, 51
　zone, 51
search-the-sky mode, 79
secondary surveillance radar (SSR), 130
　user, 157
Selective Availability (SA), 44, 68, 73
semi-major axis (SMA), 54, 138
sequencer, 80
serial tracking, 80
SGS-85 datum, 99
shuttle, 62
single-plane system, 30
slave, 133
solar gravity effect, 45

Index

sphere of position, 22, 32
spheroid, 147
sporadic-E, 51
Spot, 94
spread-spectrum, 160
Stacked Oscar On Scout program (SOOS), 146
stand-alone, 22
Standard Positioning Service (SPS), 164
Starfix, 44
sunspot, 50, 51
system performance record, 87
   zero time, 35

T
TDRSS satellite, 69
Teleglobe, 116
tetrahedron, 43
timation, 58
time lag, 22
   base, 36
   to-first-fix (TTFF), 80
TOA, 41
transfer orbit, 44
Transit, 28, 32, 143
translocation, 85
transponder, 42
transponder-assisted ranging, 130
Transverse Mercator, 149
triangle of intersection, 18
TRW, 116
Tsicada, 28

U
Universal Transverse Mercator (UTM), 148
update rate, 91
uplinked signal, 37
US Coastguard's Navigation Information Service, 72
   Federal Radionavigation Plan, 74
   Navy, 58
User Delay Range Estimate (UDRE), 179
   Equivalent Range Error (UERE), 43

V
velocity measurement, 81
Vertical Dilution of Precision (VDOP), 43
VHF communications transmitter, 78

W
water vapour, 52
wavelength, 132
weapon delivery, 58
weather satellite, 25
WGS-84, 17
Wide-Area Augmentation System (WAAS), 49, 52, 175
   WAAS Network Time (WNT), 180
Wide-area Master Station (WMS), 175
wireless time signal, 13

Y
Y code, 67